The Open
University

S342

Science: a third level course

PHYSICAL CHEMISTRY

PRINCIPLES OF CHEMICAL CHANGE

TOPIC STUDY 3
CORROSION

THE S342 COURSE TEAM

CHAIR AND GENERAL EDITOR

Kiki Warr

AUTHORS

Keith Bolton (Block 8; Topic Study 3)

Angela Chapman (Block 4)

Eleanor Crabb (Block 5; Topic Study 2)

Charlie Harding (Block 6; Topic Study 2)

Clive McKee (Block 6)

Michael Mortimer (Blocks 2, 3 and 5)

Kiki Warr (Blocks 1, 4, 7 and 8; Topic Study 1)

Ruth Williams (Block 3)

Other authors whose previous S342 contribution has been of considerable value in the preparation of this Course

Lesley Smart (Block 6)

Peter Taylor (Blocks 3 and 4)

Dr J. M. West (University of Sheffield, Topic Study 3)

COURSE MANAGER

Mike Bullivant

EDITORS

Ian Nuttall

Dick Sharp

BBC

David Jackson

Ian Thomas

GRAPHIC DESIGN

Debbie Crouch (Designer)

Howard Taylor (Graphic Artist)

Andrew Whitehead (Graphic Artist)

COURSE READER

Dr Clive McKee

COURSE ASSESSOR

Professor P. G. Ashmore (original course)

Dr David Whan (revised course)

SECRETARIAL SUPPORT

Debbie Gingell (Course Secretary)

Jenny Burrage

Margaret Careford

The Open University, Walton Hall, Milton Keynes, MK7 6AA

Copyright © 1996 The Open University. First published 1996. Reprinted 2002

Edited, designed and typeset by The Open University.

Printed in the United Kingdom by Henry Ling Ltd, The Dorset Press, Dorchester DT1 1HD

ISBN 0 7492 51921

This text forms part of an Open University Third Level Course. If you would like a copy of Studying with The Open University, please write to the Central Enquiry Service, PO Box 200, The Open University, Walton Hall, Milton Keynes, MK7 6YZ. If you have not enrolled on the Course and would like to buy this or other Open University material, please write to Open University Educational Enterprises Ltd, 12 Cofferidge Close, Stony Stratford, Milton Keynes, MK11 1BY, United Kingdom.

s342TS3i2.2

CONTENTS

1   INTRODUCTION

We are all familiar, through our domestic circumstances, with the process of corrosion — the apparent disappearance of parts of our possessions, be they cars, washing machines, central heating systems, water storage tanks, etc. In many cases we have 'cured' the corrosion problem by replacing metal systems with plastic ones; we now have plastic storage tanks and plastic guttering, for example, but for many applications there is no suitable substitute for metal.

It has been stated that corrosion is the triumph of thermodynamics over the metallurgist. Most metals are found naturally combined with other elements, such as oxygen, sulphur, etc. This means that the metals are thermodynamically unstable with respect to reaction with these elements. Expose a metal to oxygen, and in most cases we would expect some reaction to take place. But we expend a tremendous amount of resources, both human and financial, extracting metals from their ores, and in most cases we allow the metals to revert back to a thermodynamically more stable form in a very short time. Corrosion costs the nation enormous sums of money — billions of pounds each year; on a more meaningful scale, approximately one penny out of every pound spent is used in replacing corroded metals. For example, if we express corrosion in terms of the quantity of iron produced world wide, then close to 30% is used simply to replace corroded iron. This is a staggering waste of resources. Yet much of the chemistry and technology of corrosion *and its prevention* are now well understood, and it's surprising that industry and we, as members of the general public, don't know more about corrosion control.

The most familiar corrosion problem, and probably the one you are most concerned about, is corrosion of the motor car. This topic, which introduces some of the basic ideas of corrosion, and reveals how one motor manufacturer is tackling the problem, is taken up in the video sequence associated with this Topic Study (band 9 on videocassette 2, *Fighting rust in your car*).

In this Topic Study we shall introduce some thermodynamic and kinetic aspects of corrosion, which have general applicability, but we shall concentrate on the metal from which many of our everyday objects are constructed. This is the metal iron, which, as you know, is very susceptible to corrosion.

STUDY COMMENT   The components of Topic Study 3 are the main text, band 9 on videocassette 2, and an AV sequence (*The development of Pourbaix diagrams*, Section 7 of the AV booklet) that you should work through at the appropriate point in Section 2 of the main text. The video sequence acts as a good introduction to the Topic Study and could be viewed now. You will probably want to view it again before starting Section 5.

1.1   The types of reaction involved

Iron is found naturally in many different rock types — haematite, magnetite, limonite, etc. — in most of which it is combined with oxygen in various ways. Thus, iron must be thermodynamically unstable under particular conditions with respect to reaction with oxygen. It seems likely therefore that metallic iron, once extracted from its various ores, will react with the oxygen in air under normal conditions of temperature and pressure. One way of preventing compound formation — that is, corrosion — is to place the iron in an environment free from substances likely to react with it. Thus, if a piece of iron is in a vacuum it will remain unchanged almost indefinitely. However, this is not a very practical solution to the corrosion problem!

If iron is placed in an oxygen environment, oxidation of the metal occurs and various iron oxides are formed. After just one day, an oxide film approximately 1.6 nm thick is formed on the surface of the iron. However, this rate of growth rapidly diminishes: after a year, the thickness of the film is only 3.5 nm.

■ How can we explain this reduction in rate?

■ This sort of effect would be predicted if the reaction is a typical heterogeneous process in which the oxide film protects the iron surface from further reaction.

The iron exposed to oxygen under these conditions doesn't change its appearance, even after a year. Clearly, corrosion of iron by this process is not typical, but examples of it have been found. Figure 1 is a photograph of an iron pillar, now in Delhi, which was erected in the fourth century AD. The pillar is still free of the brown rust with which we are familiar, but is covered by an oxide film that gives the metal a bronzy or bluish appearance. The remarkable preservation of this pillar has been attributed to the dry (and sulfur-free) environment.

As soon as water (either as a liquid or in the vapour phase) is present in the environment of the iron, then the processes that can occur change dramatically. Under these conditions the reactions have been shown to be *electrochemical*.

1.2 The electrochemical reactions

If a drop of water (containing a small amount of sodium chloride) is placed on a horizontal sheet of iron, rust quickly appears in a roughly circular shape around the centre, as indicated in Figure 2. When ferroxyl indicator is included in the drop, then, after a short time, the drop is found to develop several rings of colour. The outer ring is pink, the next inner ring takes on the reddish-brown colour of rust, and the centre becomes blue (Figure 3). Ferroxyl indicator is a mixture of phenolphthalein and potassium ferricyanide. Potassium ferricyanide indicates the presence of iron(II) ions, since Fe^{2+} reacts with potassium ferricyanide to produce iron(II) ferricyanide — which is dark blue. Phenolphthalein is a typical acid–base indicator, which turns from colourless to pink in alkaline solutions. The experiment is demonstrated in the video sequence.

Figure 1 The iron pillar in Delhi. Legend has it that anyone who can stand, back to the column, and touch the fingers of both hands behind it, will enjoy good fortune.

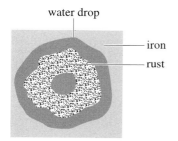

Figure 2 The formation of rust when a drop of water (containing sodium chloride) is placed on a horizontal sheet of iron.

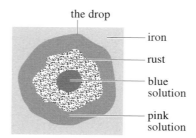

Figure 3 The formation of rust when a drop of water containing sodium chloride and ferroxyl indicator is placed on a horizontal sheet of iron.

So, can we explain what's going on inside the drop? Since the centre of the drop becomes blue, iron(II) ions are obviously being produced at this point; that is, the reaction that must be taking place is

$$Fe(s) = Fe^{2+}(aq) + 2e \tag{1}$$

At the periphery of the drop, the pink coloration suggests that the solution has become alkaline. This observation is compatible with the following reaction taking place:

$$\tfrac{1}{2}O_2(g) + H_2O(l) + 2e = 2OH^-(aq) \tag{2}$$

Thus, we have an oxidation reaction taking place at the centre of the drop, and a reduction reaction at the periphery. Since these reactions occur in different regions within the drop, the most reasonable explanation is that the reactions are taking place through an *electrochemical* process. The electrons produced by the net oxidation reaction at the anodic region (centre of the drop) are transported through the metal to the cathodic region (periphery of the drop), and the aqueous solution completes the circuit, by providing the medium for the transport of ions, and also the species required for the reduction process. However, the anode and cathode are composed of the same material, iron. Moreover, both reactions occur on the *same* piece of iron, so that electrons produced by the net oxidation reaction are *immediately* transported through the iron to the cathodic region. No external circuit is involved. Thus, the corrosion process behaves as a *short-circuited* electrochemical cell. This process is illustrated in Figure 4.

Details of why the reactions should develop in this way, with oxidation confined to the centre of the drop and reduction at the periphery, are discussed in Section 4.2. Put simply, the reduction reaction, which consumes oxygen, is confined to the periphery because in this region there is greater access to oxygen in the liquid through absorption from the surrounding air. Rust is formed when iron(II) ions formed in the oxidation process meet the hydroxide ions formed in the reduction process. Rust has a complex molecular structure, but it is commonly represented by the formula $FeO.OH$ or $Fe_2O_3.H_2O$.

The electrochemical nature of corrosion, and the effect of a difference in the concentrations of dissolved oxygen, can be demonstrated by setting up an electrochemical cell (Figure 5), in which two identical samples of iron form the anode and the cathode. If oxygen is bubbled over one electrode, a current is produced, which can be measured on the ammeter. The direction of current flow indicates that the oxygenated electrode forms the cathode. If the oxygen flow is then transferred to the other electrode, the current is reversed.

When corrosion is taking place in this way, the metal is said to be corroding by **differential aeration corrosion**.

◼   Identify a few situations in which iron will corrode via differential aeration corrosion.

▨   There are many examples of iron corroding in this way. Iron submerged in unagitated water, for example, will experience different concentrations of dissolved oxygen at different depths. The iron pillars of seaside piers will be subjected to different concentrations of oxygen at, and below, the water line. Iron pipes buried in different types of soil containing different amounts of oxygen will suffer from differential aeration corrosion. The topic is discussed in more detail in Section 4.2.

Corrosion occurring by differential aeration is an example of **heterogeneous corrosion**. The sites of the oxidation and reduction reactions are fixed, so corrosion is localized. Most corrosion tends to be localized and is therefore heterogeneous in nature, but could there be such a process as **homogeneous corrosion**? Indeed, a theory of homogeneous corrosion has been introduced to account for the corrosion of ultra-pure metals. In this type of corrosion the anodic and cathodic areas are free to move around the metal surface, and over a period of time, corrosion is uniform. Thus, the heterogeneity of the process is transitory; the areas are alternately cathodic and anodic.

Most corrosion takes place via electrochemical reactions, and we shall now restrict discussion to these reaction types. After the last two Blocks of the Course, we are now ideally poised to tackle this important topic.

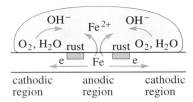

Figure 4   The corrosion process depicted as a short-circuited electrochemical cell.

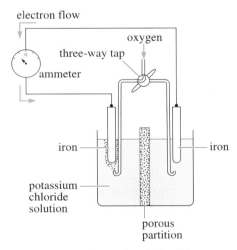

Figure 5   An electrochemical cell in which two identical samples of iron form the anode and the cathode.

1.3   Summary of Section 1

1   Most metals are thermodynamically unstable with respect to the natural environment, and so we would expect metals to corrode. Almost 30% of the iron produced world wide is used to replace corroded iron.

2   Iron can corrode by direct oxidation, but most corrosion is found to occur by electrochemical processes.

3   Examination of the corrosion that takes place when a drop of water comes into contact with iron reveals that the system behaves as a short-circuited electrochemical cell. The oxidation reaction occurs at the centre of the drop and the reduction reaction at the periphery. This is an example of differential aeration corrosion.

2 SOME THERMODYNAMIC CONSIDERATIONS:
POURBAIX DIAGRAMS

Thermodynamics tells us whether or not a particular reaction is possible, so when we are considering corrosion the crucial question is whether oxidation of the metal is possible under particular conditions. As you saw in Block 7, one approach would be to calculate $\Delta G_m^{\ominus}$ values (or to be more general, values of $dG/d\xi$) for the reaction of interest. But since we are dealing with electrochemical processes, a more direct approach is to use electrode potential data. For the reaction to be possible, E for the short-circuited electrochemical cell must be positive; that is, $E = E_{ca} - E_{an} > 0$, or $E_{ca} > E_{an}$, where the sign > is taken to mean 'more positive than', and E_{ca} and E_{an} are the electrode potentials of the half-reactions at the *cathodic* and *anodic* sites, respectively, *both written as reduction processes*.

As long as we know the reaction involved, we can use the Nernst equation to calculate the electrode potentials of the two half-reactions at particular concentrations (strictly, activities). Let's try it. Suppose we were asked the question: on thermodynamic grounds could an iron vessel hold an acid solution of pH 0 at 298.15 K without corroding? The first step would be to calculate E for a possible reduction reaction; the most likely reduction reaction, under these conditions is:

$$H^+(aq) + e = \tfrac{1}{2}H_2(g) \tag{3}$$

At pH 0 (that is, $a(H^+) = 1$), $E_{ca} = E^{\ominus}(H^+ | H_2) = 0.0\ V$.

We would then need to calculate E for a possible oxidation reaction; for example, the one mentioned in Section 1:

$$Fe(s) = Fe^{2+}(aq) + 2e \tag{1}$$

▦   Use the Nernst equation to write an expression for the *reduction* potential of this couple, E_{an}, at 298.15 K.

▦   Written as a reduction, the couple in equation 1 becomes:

$Fe^{2+}(aq) + 2e = Fe(s)$

So $E_{an} = E^{\ominus}(Fe^{2+} | Fe) - (RT/2F) \ln \{a(Fe)/a(Fe^{2+})\}$

By definition, $a(Fe) = 1$ and $a(Fe^{2+}) = \gamma c(Fe^{2+})/c^{\ominus}$. Assuming $\gamma = 1$, *as we shall throughout this Topic Study*, this equation becomes (at $T = 298.15\ K$):

$$E_{an} = E^{\ominus}(Fe^{2+} | Fe) + (RT/2F) \ln \{c(Fe^{2+})/c^{\ominus}\}$$

$$= E^{\ominus}(Fe^{2+} | Fe) + (2.303RT/2F) \log \{c(Fe^{2+})/c^{\ominus}\}$$

$$= E^{\ominus}(Fe^{2+} | Fe) + (0.029\,6\ V) \log \{c(Fe^{2+})/c^{\ominus}\}$$

But what would be the concentration of iron(II) ions to substitute into this expression? Initially, the concentration of iron(II) ions would be zero, in which case $E_{an} = E(Fe^{2+} | Fe) = -\infty$. Therefore, $E(H^+ | H_2)$ is bound to be more positive than this value (at any pH!), and the iron will start to corrode. So, according to our criterion, *any* metal in contact with *any* aqueous solution will corrode; the criterion appears to be of little use!

This problem is overcome by selecting an arbitrary, but plausible, concentration of metal ions greater than zero at which to calculate the value of E_{an}:

This concentration is $10^{-6}\ mol\ dm^{-3}$.

Thus, for the above example, $c(Fe^{2+}) = 10^{-6}$ mol dm^{-3}, $c^{\ominus} = 1$ mol dm^{-3} and $E^{\ominus}(Fe^{2+}\,|\,Fe) = -0.46$ V (S342 *Data Book*), so:

$$E_{an} = [-0.46 + 0.029\,6(-6)] \text{ V}$$

$$= -0.46 \text{ V} - 0.178 \text{ V}$$

$$\approx -0.64 \text{ V}$$

Since $E_{ca} > E_{an}$, the process will take place on thermodynamic grounds: the iron will corrode.

The above calculation was based on the choice of one particular pair of reactions, but there are other possibilities. For example, the reduction reaction could be the one discussed earlier:

$$\tfrac{1}{2}O_2(g) + H_2O(l) + 2e = 2OH^-(aq) \tag{2}$$

Two other possible oxidation reactions are

$$Fe(s) = Fe^{3+}(aq) + 3e \tag{4}$$

$$3Fe(s) + 4H_2O(l) = Fe_3O_4(s) + 8H^+(aq) + 8e \tag{5}$$

Indeed, there is a whole range of other possibilities. In fact, the calculation of E_{an} values for all the possible reactions involving iron is particularly complicated, because iron has so many oxidation states and so many different oxides. In addition, the thermodynamic feasibility of many of these processes depends crucially on the pH of the solution. So, instead of calculating the E_{an} value for every pH condition, we can summarize the information on a plot of V versus pH for some likely reactions. This plot is called a **Pourbaix diagram** (after the French electrochemist M. J. N. Pourbaix). A simplified Pourbaix diagram for iron is shown in Figure 6.

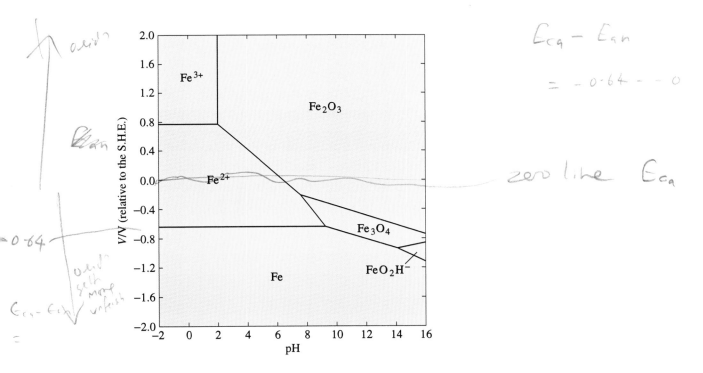

Figure 6   A simplified Pourbaix diagram for iron.

STUDY COMMENT   Part of this diagram for iron is developed in the tape sequence associated with this Topic Study. You should now turn to Section 7 in the AV Booklet and work through this material. It should take you about 45 minutes in all.

You should now be familiar with Pourbaix diagrams, both in their construction and in their interpretation. But what do these diagrams tell us about the *corrosion* characteristics of a particular metal? Well, the Pourbaix diagram shown in Figure 6, for example, reveals that iron is stable at large negative potentials (relative to the standard hydrogen electrode, S.H.E.). Under these conditions a piece of iron will remain as iron metal and not corrode. Thus, the Fe region in the Pourbaix diagram can be labelled '**immunity**': the iron is immune to corrosion under these conditions. Above the Fe region is the Fe^{2+} region; at any point in this region, iron will be converted into iron(II) ions, so according to thermodynamic arguments the iron will corrode. This region can therefore be labelled as 'corrosion'. Similarly, in the Fe^{3+} region, iron will tend to form Fe^{3+} ions, and this region can therefore also be labelled as 'corrosion'. In the right-hand regions of the Pourbaix diagram the oxides Fe_2O_3 and Fe_3O_4 are formed. These are solid compounds, and it is possible that these oxides may form a coating on the metal surface, protecting the metal from further attack by water and oxygen. The formation of such a protective coating is called **passivation**.

With all this in mind, we can now represent the corrosion behaviour of iron, at 25 °C, by the diagram shown in Figure 7 (referred to as a **Pourbaix corrosion diagram**). Similar diagrams for zinc and copper are shown in Figures 8 and 9, respectively.

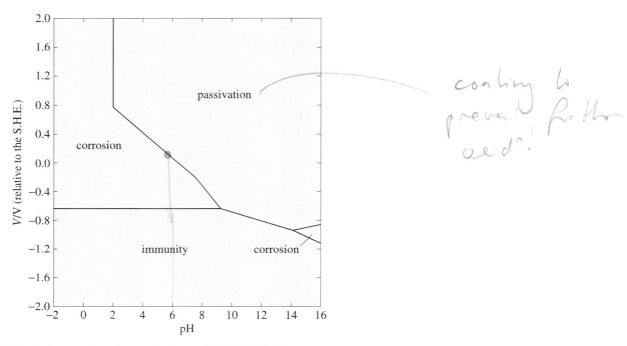

Handwritten annotations:

$E_{cell} = -ve$

$\therefore \Delta G = +ve$

coating to prevent further ox'd^n.

Figure 7   The Pourbaix corrosion diagram for iron at 298.15 K (25 °C).

■   By reference to Figure 7, suggest two ways of protecting iron from corrosion.

■   Figure 7 reveals that the corrosion of iron can be prevented by manipulation of the pH and the value of the potential difference, *V*. If the potential difference is made large and negative (relative to the S.H.E.) at any pH value, then the iron falls into the immunity region of the corrosion diagram. Thus, iron *cannot* rust under these conditions. Similarly, if the *V* and pH values are adjusted such that the iron is in the passivation region of the corrosion diagram, the rusting of iron can be considerably reduced.

In practice, the degree of passivation depends on the amount of protection afforded by the oxide layers. To be protective the oxide film must be mechanically stable, so that it doesn't flake or crack; it must be non-porous, and it must prevent ionic conduction.

The manipulation of pH is straightforward, but how do we manipulate the value of *V*? In practice, this is achieved by connecting the corroding system to some external

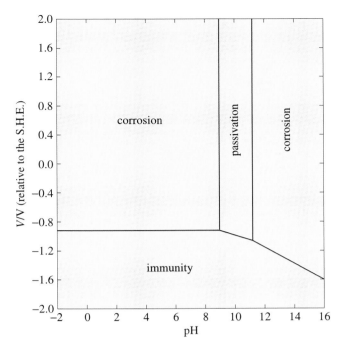

Figure 8 The Pourbaix corrosion diagram for zinc.

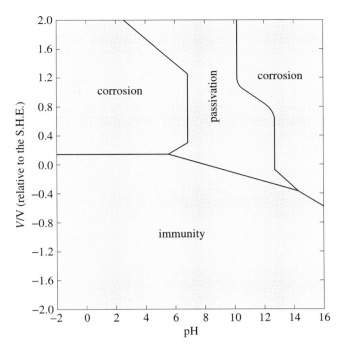

Figure 9 The Pourbaix corrosion diagram for copper.

voltage source or by connecting the metal to a so-called 'sacrificial metal': a metal more prone to corrosion than the corroding system. This topic is taken up later (in Section 5.2). However, this is not as simple an answer to the corrosion problem as it might, at first, appear!

With these external devices we can manipulate the potential difference across the interface between the metal and the solution, and so control corrosion. But what about the more normal situation where devices like these are not present? Suppose we have a piece of iron, say, in a particular environment: it is not immediately obvious how the Pourbaix diagram will help us decide whether or not corrosion is possible. The problem is that, so far, we have considered only the possible anodic reactions. To find out what *overall* reactions are possible we must also consider the likely cathodic reactions for water in contact with iron. There are two important cathodic reactions to consider, both of which we've met before:

$$H^+(aq) + e = \tfrac{1}{2}H_2(g) \tag{3}$$

and

$$\tfrac{1}{2}O_2(g) + H_2O(l) + 2e = 2OH^-(aq) \tag{2}$$

These can be incorporated into the Pourbaix diagram.

- On the Pourbaix diagram, will the lines for these two reactions be horizontal, vertical or sloping?

- Since both reactions involve electrons *and* either H^+ or OH^- ions, the lines will slope.

The lines for these two reactions are superimposed on the Pourbaix corrosion diagram for iron in Figure 10, and you can see that the oxygen cathodic line lies near the top of the figure. This means that the oxygen cathodic reaction can couple with any of the anodic reactions presented below it (since $E_{ca} > E_{an}$) to yield a positive value of E — that is, a thermodynamically possible situation. Thus, when iron is in contact with oxygenated water, it can fall into any of the regions below the oxygen cathodic line. The actual reaction that takes place depends on the pH of the solution. At low pH, corrosion takes place; at higher pH, passivation of the metal occurs.

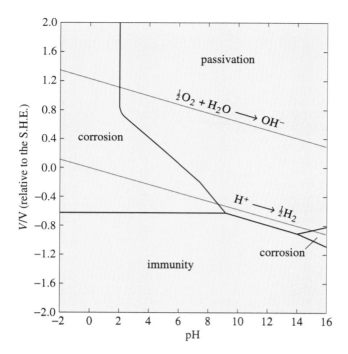

Figure 10   The Pourbaix corrosion diagram for iron, including likely cathodic processes (green lines).

Figure 10 also reveals that if the cathodic reaction is formation of hydrogen gas, corrosion of iron is possible over most values of pH. However, on thermodynamic grounds we would expect that the most probable cathodic reaction would be the one that provides the most negative value of $dG/d\xi$ (that is, the one that produces the most positive value of E). According to Figure 10, this is likely to be the oxygen reduction reaction, at all values of pH. But, as we shall see later (Section 3.2), this reaction is slow, and conditions can be such that hydrogen gas formation becomes the more probable reaction.

As you might expect, changing the conditions alters the corrosion diagram. For example, increasing the temperature usually increases the extent of the ionic regions, thereby increasing the extent of the corrosion domain. Even more dramatic changes can be brought about by the addition of other ions. Thus, the addition of chloride ions (which reflects more closely the conditions present in seawater, for example) changes the iron corrosion diagram to that shown in Figure 11. Comparison of Figure 11 with Figure 10 reveals that the corrosion domain is increased substantially. This topic is discussed further in Section 4.4.

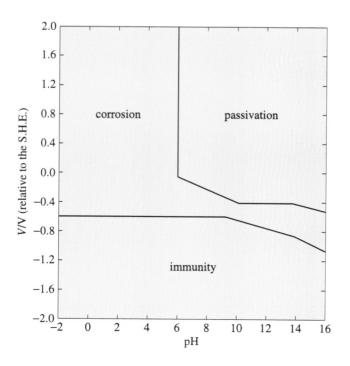

Figure 11   The Pourbaix corrosion diagram for iron in the presence of chloride ions.

These types of Pourbaix corrosion diagram, based on thermodynamic data, are extremely useful in studying the corrosion behaviour of various metals. But as with all arguments based solely on thermodynamic considerations, predictions drawn from the diagrams must be interpreted with caution.

SAQ I   The Pourbaix corrosion diagrams shown in Figure 12 are for four metals — gold, rhodium, aluminium and hafnium. You are given the following information, based on thermodynamic data: (i) rhodium is less susceptible to corrosion than gold; (ii) hafnium is less susceptible to corrosion than aluminium; (iii) aluminium cooking utensils can be safely used around pH 7.

Assuming that susceptibility to corrosion is based on the size of the appropriate regions shown, (a) decide which diagram corresponds to which metal, and (b) label the corrosion and/or passivation regions in each diagram.

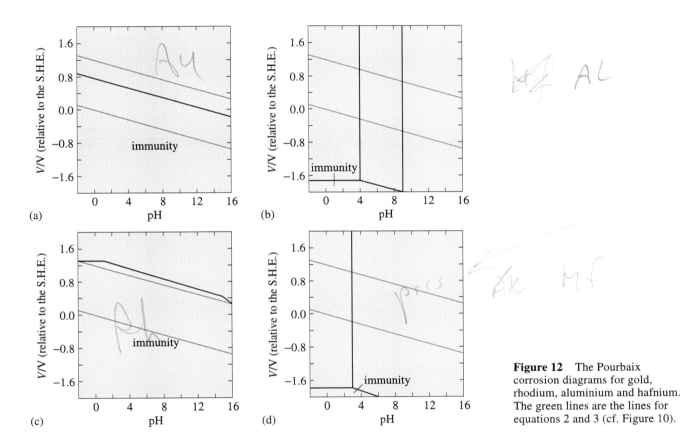

Figure 12   The Pourbaix corrosion diagrams for gold, rhodium, aluminium and hafnium. The green lines are the lines for equations 2 and 3 (cf. Figure 10).

2.1   Summary of Section 2

1   On thermodynamic grounds, a metal is said to corrode if E_{ca} is more positive than E_{an} when the concentration of the metal ion is put (somewhat arbitrarily) at 10^{-6} mol dm^{-3}.

2   The E_{an} information for a metal can be conveniently summarized in a Pourbaix diagram, which shows the regions of stability of various forms of the metal in different oxidation states, as a function of pH.

3   The Pourbaix diagram can be converted into a Pourbaix corrosion diagram by assuming that: regions in which aqueous ions are stable yield corrosion domains; regions where solid oxidation products are formed yield passivation domains; and regions in which the pure metal is stable yield immunity domains.

4   In order to see if corrosion is possible on thermodynamic grounds, the various likely cathodic reactions must be superimposed on the Pourbaix corrosion diagram. Regions falling below the line of the cathodic reaction are possible on thermodynamic grounds.

5   The Pourbaix corrosion diagram for iron (in the presence of oxygen and water) reveals that corrosion of iron is possible on thermodynamic grounds if the cathodic reaction is reduction of oxygen gas or production of hydrogen gas.

6   The corrosion diagram is altered by the addition of other ions, or by changes in temperature.

3   SOME KINETIC CONSIDERATIONS

Thermodynamic data provide information on whether or not a particular reaction is feasible, allowing us to construct corrosion diagrams that show the conditions necessary for corrosion, passivation and immunity. But although conditions may be such that iron should be in the corrosion region, say, this doesn't necessarily mean that corrosion will occur. The anodic reaction leading to corrosion could be very slow. Similarly, saying that under particular conditions iron will be protected from corrosion by the formation of insoluble and impervious oxide layers will be true *in practice* only if the kinetics of the processes involved are reasonable. Thus, we must now extend our discussion to consider some kinetic aspects of corrosion.

During corrosion, the system is behaving as a short-circuited electrochemical cell. Suppose, for simplicity, that the reactions involved are as follows:

anodic: $Fe(s) = Fe^{2+}(aq) + 2e$ $\qquad\qquad$ (1)

cathodic: $H^+(aq) + e = \tfrac{1}{2}H_2(g)$ $\qquad\qquad$ (3)

The electrochemical cell that is involved can be expressed as

$Fe(s)|\,Fe^{2+}(aq),\,H^+(aq)\,|\,H_2(g),\,Fe(s)$

Notice that the cell is a very simple one: in particular, there is just *one* solution and no connections to an external circuit, so there are only two interfaces to consider.

■   Write down an expression for the potential difference operating across this cell, V_{cell}.

▨   Reading from the right (as we did in Block 8):

$$V_{cell} = {}^{Fe,\,H_2}\Delta^{H^+}\phi + {}^{Fe^{2+}}\Delta^{Fe}\phi + IR_S \qquad\qquad (6)$$

where I is the current flowing and R_S is the internal resistance of the cell.

But, in corrosion, the cell is short-circuited, and therefore $V_{cell} = 0$. Thus,

$${}^{Fe,\,H_2}\Delta^{H^+}\phi + {}^{Fe^{2+}}\Delta^{Fe}\phi + IR_S = 0 \qquad\qquad (7)$$

If the solution in contact with the metal is conducting, and if the site of the cathodic reaction is physically close to the site of the anodic reaction, then it seems reasonable to suggest that $IR_S \approx 0$. Under these circumstances, equation 7 becomes

$${}^{Fe,\,H_2}\Delta^{H^+}\phi = -{}^{Fe^{2+}}\Delta^{Fe}\phi$$

$$= {}^{Fe}\Delta^{Fe^{2+}}\phi \qquad\qquad (8)$$

We conclude that *the potential differences across the two metal–solution interfaces are the same*. This potential difference is called the **corrosion potential difference**, and is given the symbol $\Delta\phi_{corr}$. In this case,

$$\Delta\phi_{corr} = {}^{Fe}\Delta^{Fe^{2+}}\phi = {}^{Fe, H_2}\Delta^{H^+}\phi \qquad (9)$$

or more generally

$$\Delta\phi_{an} = \Delta\phi_{ca} = \Delta\phi_{corr} \qquad (10)$$

So we now have an expression for the potential difference at which corrosion is occurring, but what about the *rate* of corrosion? Well, the rate of corrosion can be expressed as the rate of disappearance of the metal. But the rate of disappearance of the metal is related to the rate of production of electrons via the anodic reaction, which is simply measured as I_{an}, the anodic current. Thus,

$$\text{corrosion rate} = -\frac{d[Fe]}{dt} \propto \frac{d[e]}{dt} = I_{an} \qquad (11)$$

Thus, the corrosion rate, expressed as a *current* — the **corrosion current, I_{corr}** — is equal to I_{an}. But the rate of production of electrons via the anodic reaction must, under steady-state conditions, be equal to the rate of consumption of electrons via the cathodic reaction. Thus, *the anodic current must be equal in magnitude, but opposite in sign to the cathodic current.* That is,

$$I_{corr} = I_{an} = -I_{ca} \qquad (12)$$

Thus, to summarize, when corrosion is taking place, the potential differences across the metal–solution interfaces are the same, and the rate of corrosion is determined when the anodic and cathodic currents are equal. We shall now use this information to construct diagrams of potential difference versus I (or log I) for a corroding system. Such diagrams are called *Evans diagrams*. In their simplest form, they are a combination of experimental Tafel plots for the cathodic and anodic processes.

3.1 Evans diagrams

Consider first of all the iron oxidation reaction

$$Fe(s) = Fe^{2+}(aq) + 2e \qquad (1)$$

■ What does a plot of current *density*, i, versus overpotential, η, look like for this sort of process?

▨ Plots of i versus η were introduced in Block 8 (Section 3.2). For the reaction taking place at a single electrode, a plot of i versus η, at relatively small values of η, will probably resemble Figure 13.

■ Which part of this curve do we require in order to represent the corrosion process?

▨ If corrosion is taking place, then the metal is undergoing *net* oxidation, which is represented by the top-right section of the plot in Figure 13.

■ What about the coupled process taking place at the cathodic site? How could we represent this?

▨ For the process taking place at the cathodic site we would need to construct another i versus η curve, as indicated in Figure 14. Although this would probably have the same general features as the plot in Figure 13, it could, of course, be very different in detail. To represent the cathodic, net reduction reaction, we need to consider only the bottom-left section of the plot in Figure 14.

So the two processes taking place can be represented by the top-right section of Figure 13 and the bottom-left section of Figure 14. Can we combine these two plots

Figure 13 A possible plot of i versus η for the iron oxidation reaction.

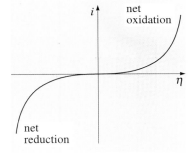

Figure 14 A possible plot of i versus η for the cathodic process.

in some way? Indeed, we can! The argument is similar to the one we developed in order to discuss competing electron-transfer processes at a single electrode (Block 8, Section 5.3). As there, the secret is to replace the η term on the horizontal axis by a term involving the actual potential difference across the interface, given the following expressions:

$$\eta_{an} = \Delta\phi_{an} - \Delta\phi_{e,an} \tag{13}$$

difference between + and equilibrium

(where 'an' represents the net oxidation process at the anodic site, and ϕ_e is the equilibrium potential difference), and

$$\eta_{ca} = \Delta\phi_{ca} - \Delta\phi_{e,ca} \tag{14}$$

(where 'ca' represents the net reduction process at the cathodic site).

But, for the special case of corrosion, we have shown that

$$\Delta\phi_{an} = \Delta\phi_{ca} = \Delta\phi_{corr} \tag{10}$$

So, by subtracting equation 14 from equation 13, it follows that:

$$\eta_{an} - \eta_{ca} = \Delta\phi_{e,ca} - \Delta\phi_{e,an} \tag{15}$$

This expression effectively tells us that the two curves on a plot of i versus $\Delta\phi$ will be separated along the $\Delta\phi$ axis by the difference between the values of their equilibrium potential differences, as indicated in Figure 15.

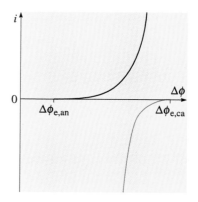

Figure 15 Plots of i versus $\Delta\phi$ for both the anodic and cathodic processes.

- ■ Figure 15 is drawn with $\Delta\phi_{e,ca}$ more positive than $\Delta\phi_{e,an}$: is this necessarily so?

- ■ Yes. Since, $\eta_{an} > 0$ and $\eta_{ca} < 0$ (recall Block 8, Section 6.4), $\eta_{an} - \eta_{ca}$ must be positive. Hence (from equation 15), $\Delta\phi_{e,ca}$ must be more positive than $\Delta\phi_{e,an}$.

Thus, at any value of the potential difference across the metal–solution interface we can determine i_{an} and i_{ca}. However, we know that when corrosion is taking place the net *currents* for oxidation and reduction are equal in magnitude, $I_{an} = -I_{ca}$. So, instead of plotting current density, it is more useful to plot current. This can be done by multiplying the current density by the area of the site at which each process is occurring. If the areas are similar, the resulting plot, Figure 16, will have the same general appearance as Figure 15.

But the point at which the currents are equal is determined more easily if the magnitude of I, $|I|$, rather than the current itself is plotted — as shown in Figure 17. As indicated, the point of intersection of the two curves gives the corrosion current, I_{corr}, and the corrosion potential, $\Delta\phi_{corr}$. However, this is still not an Evans diagram! To produce an Evans diagram we need to incorporate two more changes.

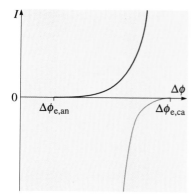

Figure 16 Plots of I versus $\Delta\phi$ for both the anodic and cathodic processes.

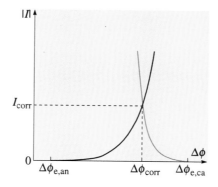

Figure 17 Plots of $|I|$ versus $\Delta\phi$ for both the anodic and cathodic processes.

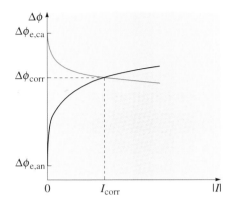
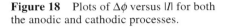

Figure 18 Plots of $\Delta\phi$ versus $|I|$ for both the anodic and cathodic processes.

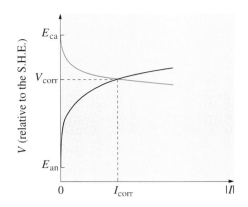

Figure 19 An Evans diagram: a plot of V (relative to the S.H.E.) versus $|I|$ for both the anodic and cathodic processes. The point of intersection of the two curves gives the corrosion current, I_{corr}, and the corrosion potential, V_{corr} (relative to the S.H.E.).

First, we need to transpose the axes (Figure 18) and then equate the $\Delta\phi_e$ values (which, as we discussed in Block 8, can't be determined) with the electrode (reduction) potential values for the two half-reactions in question. In other words, we once again use the following *approximations*:

$$\Delta\phi_{e,ca} \approx E_{ca} \quad \text{and} \quad \Delta\phi_{e,an} \approx E_{an} \tag{16}$$

and hope that this doesn't introduce too much error. The vertical axis can now be labelled V, the potential difference of the system, relative to the S.H.E. The resulting diagram, shown in Figure 19, is called an **Evans diagram**. For the purposes of this Topic Study, an alternative display — a plot of V versus $\log |I|$ (a version of which is shown in Figure 20) — will, in many cases, be more useful.

In constructing these plots for real systems we have a problem. As we said in Section 1.2, most corrosion occurs via a heterogeneous process: corrosion tends to be localized. In these situations it is often extremely difficult to know the relative areas of the cathodic and anodic regions, and so calculation of values of the exchange currents from the exchange current densities becomes difficult. In many of the examples that follow it will be assumed that the area of the cathodic site is equal to that of the anodic site, although this may be far from true. Of course, if corrosion occurs via a homogeneous mechanism, then this assumption is perfectly justifiable; but this is rarely so.

When the areas of the anodic and cathodic sites can be equated, Figure 20 can be redrawn with the current density rather than the current on the horizontal axis. The intersection of the two lines then gives an idea of the *intensity* of corrosion; it gives the value of the corrosion current density. This is indicated in Figure 21, which reveals that the rate of corrosion, expressed as a **corrosion current density**, depends on many quantities.

■ Try to list some of these quantities.

▪ The corrosion current density depends on (a) the difference betweeen the E values of the two processes; (b) the Tafel slopes (that is, the values of the transfer coefficients, α, for the two processes); (c) the values of the exchange current densities.

■ Changes in concentration are obviously going to affect the rate of corrosion. Do concentration terms appear in the Evans diagram?

▪ Changes in concentration manifest themselves as changes in the values of E, and also in the values of the exchange current densities.

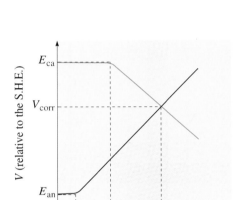

Figure 20 Another version of an Evans diagram — a plot of V (relative to the S.H.E.) versus $\log |I|$. $I_{e,an}$ is the exchange current for the anodic process, and $I_{e,ca}$ is the exchange current for the cathodic process.

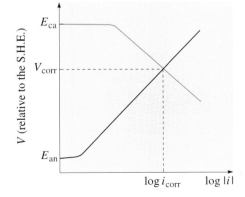

Figure 21 A third version of an Evans diagram — a plot of V (relative to the S.H.E.) versus $\log |i|$. The point of intersection gives the log of the corrosion current density.

You should now attempt SAQ 2.

SAQ 2 With reference to Figure 21, and assuming that values for the anodic reaction remain constant, sketch graphs of V (relative to the S.H.E.) versus $\log |i|$ to show how the rate of corrosion is affected by: (i) an increase in the value of E_{ca} (that is, to more positive values); (ii) an increase in the value of $i_{e, ca}$; (iii) an increase in the value of α_{red} (strictly, $\alpha_{ca, red}$ using the notation from Block 8) for the cathodic reaction.

Apart from the changes discussed in the answer to SAQ 2, Evans diagrams can also be used to show the effect on the rate of corrosion if the rate of one of the processes becomes limited by concentration polarization, as indicated in Figure 22.

One final point: if the condition that $IR_S = 0$, used in the construction of all the above Evans diagrams, doesn't hold, then this also can be indicated in the Evans diagram. The curves no longer intersect at the value of the corrosion current density, but are separated at this point by the IR_S term (see Figure 23, where $IR_S = I_{corr}R_S$).

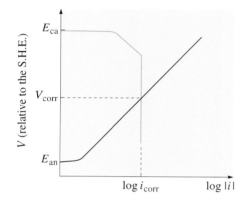

Figure 22 An Evans diagram in which the cathodic process is limited by concentration polarization.

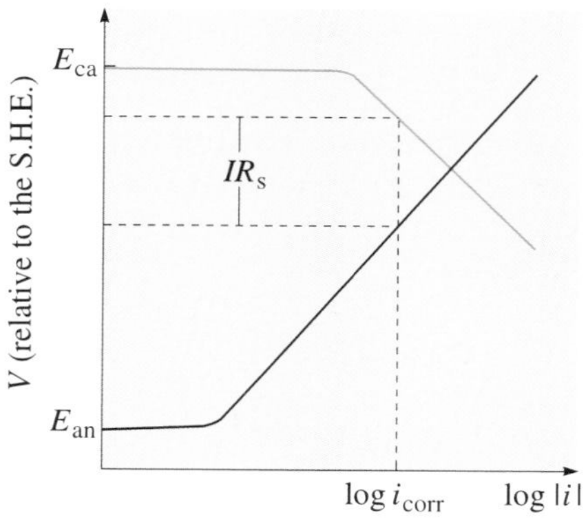

Figure 23 An Evans diagram in which $IR_S \neq 0$.

You should now try SAQ 3, which brings together many of the important ideas and concepts developed so far.

SAQ 3 Construct an Evans diagram to calculate the corrosion current density for zinc under acid (pH 2) conditions (at 298.15 K). Obtain values of $E^{\ominus}$ from the S342 *Data Book*.

Assume that:

(a) the cathodic and anodic areas are equal;

(b) $IR_S = 0$;

(c) the process is not limited by concentration polarization;

(d) the cathodic reaction is reduction of hydrogen ions;

(e) the transfer coefficient for the reduction of hydrogen ions is 0.5, and the transfer coefficient for the oxidation of zinc metal is 1.5;

(f) i_e values under corrosion conditions are the same as those under standard concentration conditions; that is, 10^{-6} A m^{-2} for the formation of hydrogen on zinc, and 10^{-3} A m^{-2} for the formation of zinc on zinc (from Block 8, Table 1);

(g) $c(Zn^{2+}) = 10^{-6}$ mol dm^{-3}.

Which of the assumptions (a)–(g) is likely to be least valid?

3.2   Which is the cathodic reaction?

We have already stated that there are typically two possible cathodic processes that can couple with the metal anodic process,

either $H^+(aq) + e = \frac{1}{2}H_2(g)$ (3)

or $\frac{1}{2}O_2(g) + H_2O(l) + 2e = 2OH^-(aq)$ (2)

According to thermodynamic considerations, it should be the oxygen cathodic reaction that predominates, since this yields the larger value of E, whatever the value of the pH (see the Pourbaix corrosion diagram for iron, Figure 10). However, under acidic conditions it is the hydrogen cathodic reaction that usually occurs. Indeed, this reaction was the assumed reaction in SAQ 3. We shall now use Evans diagrams to explain this apparent discrepancy, using iron corrosion as an example.

Let us construct Evans diagrams at two pH values, say pH 0 and pH 6. (To go any higher in pH takes the system into regions where the mechanisms of the reactions are not fully documented: in addition, at a pH higher than about 8, the Pourbaix diagram for iron (Figure 6) reveals that various iron oxides are formed, and this complicates the story. Such complications are taken up in Section 3.3.)

Although the iron oxidation reaction appears straightforward, this is not so. As we established earlier, if $c(Fe^{2+}) = 10^{-6}$ mol dm^{-3}, then $E(Fe^{2+}|Fe) = -0.64$ V. The mechanism of the iron oxidation reaction was discussed in Block 8 (SAQ 4). There, the transfer coefficient for the *oxidation* process was said to be 1.5, which would produce a slope to the Tafel line of 40 mV. The value of i_e for the reaction, under standard concentration conditions (at pH 0), is given as 10^{-4} A m^{-2} (Block 8, Table 1). Rather than assume, as we did in SAQ 3, that the value of i_e is effectively constant, let us now take the change in concentration (from 1.0 to 10^{-6} mol dm^{-3}) into account. In this case, experiments have shown that the value of i_e is proportional to $c(Fe^{2+})$, so a decrease in the concentration of iron(II) ions by 10^6 decreases the value of i_e by 10^6. Thus, we would expect the value of i_e for corroding iron to be $10^{-4} \times 10^{-6}$ A m^{-2} = 10^{-10} A m^{-2}. But will this value of i_e be affected by the pH of the solution? At first sight we might answer no to this question, but there is a complication. Examination of the mechanism of the iron oxidation reaction (Block 8, SAQ 4) reveals that it involves hydrogen ions, and so the value of i_e could be dependent on the pH of the solution. In fact, experiments have shown that the value of i_e is proportional to $c(OH^-)$. This means that an increase in the concentration of hydroxide ions by 10^6 increases the value of i_e by 10^6.

■   What then, are the values of i_e under corrosion conditions, at pH 0 and at pH 6?

▨   At pH 0 the value of i_e will be 10^{-10} A m^{-2} (as calculated above). At pH 6 the concentration of OH$^-$ has been increased by a factor of 10^6, so the value of i_e will be $10^{-10} \times 10^6$ A m^{-2} = 10^{-4} A m^{-2}.

Let us now assume that the areas of the anodic and cathodic sites are equal, so that the Evans diagram can be constructed with current density, rather than current, plotted along the horizontal axis. If two such Evans diagrams are drawn, one at pH 0 (Figure 24a) and one at pH 6 (Figure 24b), then the Tafel line for the iron oxidation process can be inserted into each diagram.

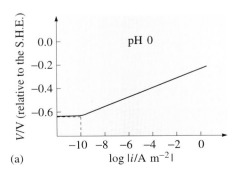

(a)

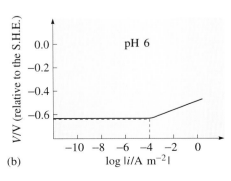

(b)

Figure 24   Part of the Evans diagram for iron (a) at pH 0, and (b) at pH 6. In both cases the Tafel slope is 40 mV.

Now let's turn to consider the possible cathodic reactions. At pH 0:

$$E(H^+ | H_2) = E^{\ominus}(H^+ | H_2) = 0.0 \text{ V}$$

The exchange current density for the reduction of hydrogen ions on iron is given in Table 1 of Block 8 as 10^{-2} A m^{-2}, under standard concentration conditions. A pH value of zero corresponds to a standard concentration (strictly, activity) of hydrogen ions, so the value of i_e will be 10^{-2} A m^{-2} at pH 0.

The alternative cathodic reaction was discussed at several points in Block 8. In particular, you saw that it can be written in two superficially different, but effectively equivalent, forms, depending on the pH of the solution: either as equation 2 (for high pH values):

$$\tfrac{1}{2}O_2(g) + H_2O(l) + 2e = 2OH^-(aq) \tag{2}$$

or as follows (for low, or acidic, pH values):

$$\tfrac{1}{2}O_2(g) + 2H^+(aq) + 2e = H_2O(l) \tag{17}$$

As you saw in Block 8, application of the Nernst equation reveals that the pH dependence of the emf is identical for these two forms. Thus, to take the more familiar acidic form (equation 17),

$$E(17) = E^{\ominus}(17) - (0.059\,2 \text{ V}) \text{ pH}$$

$$= E^{\ominus}(17) \text{ at pH 0}$$

$$= 1.23 \text{ V}$$

Equation 17 is also the form quoted in Table 1 of Block 8: in this case, pH 0 again corresponds to standard concentration conditions, so i_e will have the value listed there; that is, 10^{-10} A m^{-2}.

For both cathodic processes the value of the transfer coefficient is 0.5, which means that the Tafel slopes will be -120 mV. Thus, Tafel lines for the two possible cathodic processes can now be added to Figure 24a, to produce Figure 25.

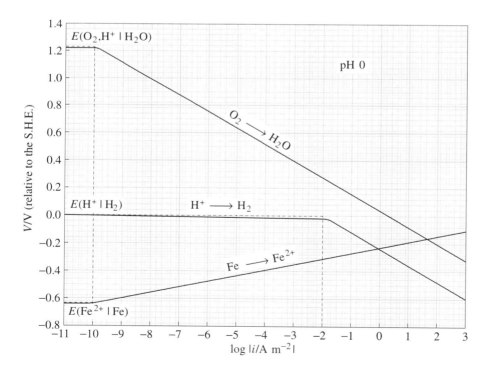

Figure 25 The Evans diagram for iron at pH 0, including both possible cathodic processes.

■ According to Figure 25, which is the more likely cathodic process?

▨ Figure 25 reveals that at pH 0 the greater corrosion current density will be produced if the iron anodic reaction couples with the *oxygen cathodic* reaction. But this is not what happens in practice.

■   Where might our interpretation of the Evans diagram be in error?

▪   For the intersection of the two Tafel lines to reveal the value of the corrosion current density, (a) the cathodic and anodic areas must be equal, (b) the value of IR_S must be zero, and (c) the reactions should not be restricted by concentration polarization.

As you might expect, the oxygen cathodic reaction is *very* susceptible to concentration polarization, since the rate of the reaction will often be controlled by the rate of diffusion of dissolved oxygen in the liquid phase. In fact, the limiting current density for oxygen reduction is of the order of $0.1\ A\ m^{-2}$ in *unagitated* solutions. Thus, the oxygen cathodic line shown in Figure 25 should be discontinued when the current density has reached a value of $0.1\ A\ m^{-2}$. This additional information means that Figure 25 should be modified to the form shown in Figure 26.

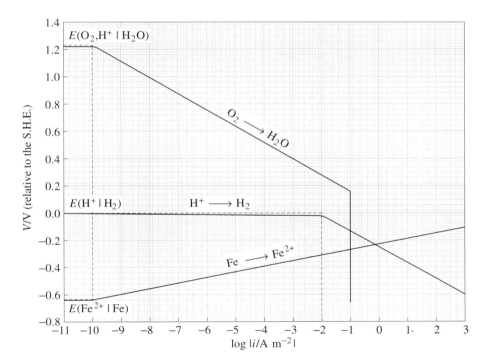

Figure 26   The Evans diagram for iron at pH 0, showing that the oxygen cathodic process is subject to concentration polarization at a value of i of $0.1\ A\ m^{-2}$ in unagitated solutions.

Figure 26 reveals that the corrosion current density produced by the hydrogen cathodic reaction is now in excess of that produced by the oxygen cathodic reaction, so hydrogen ion reduction becomes the predominant process. You should note that in calculating the E value for the oxygen cathodic reaction, we also implicitly assumed a partial pressure of oxygen gas of 1 bar. But the partial pressure of oxygen above the solution is likely to be less than this. As you may like to check for yourself, this would reduce the value of E, further increasing the probability of the hydrogen cathodic reaction. Of course, once the hydrogen cathodic reaction has started and the iron has begun to corrode, the hydrogen gas produced will tend to flush any oxygen out of the solution, so making oxygen reduction even less likely.

Figure 26 depicts the situation in unagitated solutions at pH 0. If the solution is stirred, however, the picture changes: the value of the limiting current density is increased, and, at low pH values, the oxygen cathodic reaction becomes predominant. Similarly, if oxygen gas is pumped into an aqueous solution, the Evans diagram shown in Figure 25 may be a better representation of the experimental situation: the rate of corrosion, driven by the oxygen cathodic reaction, may then be substantial. For example, Figure 25 reveals that the rate of corrosion of iron in oxygen-saturated acid can be over ten times the rate in hydrogen-saturated acid (that is, a corrosion current density of $10^{1.5}\ A\ m^{-2}$ rather than $1\ A\ m^{-2}$); this is borne out in practice.

To complete the analysis, let us now repeat the calculations at pH 6. For the hydrogen cathodic process:

$$E(\mathrm{H^+ \,|\, H_2}) = (-0.059\,2\ \mathrm{V})\,\mathrm{pH} = -0.36\ \mathrm{V\ at\ pH\,6}$$

As before, the exchange current density, at unit activity of hydrogen ions, has a value of $10^{-2}\ \mathrm{A\,m^{-2}}$. However, the value of i_e is a function of $c(\mathrm{H^+})^{1/2}$, so a *decrease* in the hydrogen ion concentration of 10^6 (from pH 0 to pH 6) decreases the value of i_e by 10^3. So we estimate that the value of i_e for the reduction of hydrogen ions on iron at pH 6 is $10^{-5}\ \mathrm{A\,m^{-2}}$.

For the alternative cathode reaction (equation 17):

$$E(17) = 1.23\ \mathrm{V} - (0.059\,2\ \mathrm{V})\,\mathrm{pH} = 0.87\ \mathrm{V\ at\ pH\,6}$$

Our discussion so far suggests that the value of i_e will be different from that under standard concentration conditions ($10^{-10}\ \mathrm{A\,m^{-2}}$). We can estimate the value of i_e at pH 6 if we assume (and it's a big assumption) that it has the same dependence on $c(\mathrm{H^+})$ as does the hydrogen ion reduction reaction; that is, we assume that i_e is proportional to $c(\mathrm{H^+})^{1/2}$. Thus, at pH 6 the value of i_e will be $10^{-10} \times (10^{-6})^{1/2} = 10^{-13}\ \mathrm{A\,m^{-2}}$. Unfortunately, we are obliged to make assumptions like this because so few data are available on these types of reaction. Assuming also that the mechanisms of the cathodic reactions are unchanged on moving from pH 0 to pH 6, we can again put the transfer coefficients for both reactions at 0.5. These Tafel lines can be added to Figure 24b to produce Figure 27.

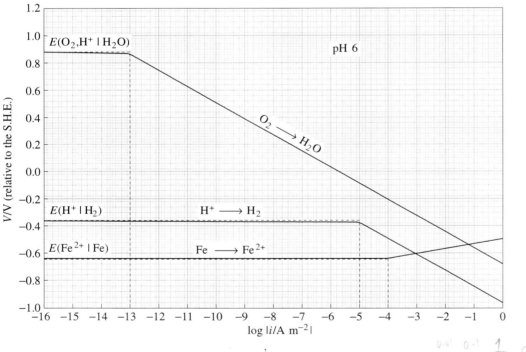

Figure 27 The Evans diagram for iron at pH 6, including both possible cathodic processes.

Figure 27 reveals that the larger current density results from the coupling of the iron anodic reaction with the *oxygen* cathodic reaction. Because the corrosion current density at this pH value is less (but only just!) than the limiting current density of the oxygen cathodic reaction, no complications occur in this example. Thus, as the pH changes, we expect the nature of the cathodic reaction to change; under normal (unagitated) conditions, the hydrogen cathodic reaction predominates at low pH values, whereas the oxygen cathodic reaction predominates at higher pH values. If the pH is raised even further, various iron oxides and hydroxides are formed, which cover the surface of the iron and protect it from further corrosion (cf. Figures 6 and 7).

How do our corrosion current densities calculated from Evans diagrams compare with those observed in practice? Comparison of Figure 27 with Figure 26 reveals that the rate of corrosion of iron is reduced considerably on moving to a higher pH value. At pH 0, the corrosion current density is of the order of $1.0 \, \text{A m}^{-2}$ (in unagitated solutions), whereas it falls to around $10^{-1.2} \, \text{A m}^{-2}$ at pH 6, a reduction in corrosion rate by a factor of 16. This *change* in rate compares favourably with the experimental data because the rate of corrosion of iron at pH 0, in unagitated solutions, is approximately 30 mm per year, whereas at pH 6 it is approximately 1 mm per year; that is, a reduction by a factor of 30. Comparison of the calculated value *at a particular value of pH* is less favourable. At pH 0, an observed corrosion rate of 30 mm per year can be converted into a corrosion current density value of about $30 \, \text{A m}^{-2}$, which should be compared with the calculated value of $1 \, \text{A m}^{-2}$. However, the agreement is reasonable, taking into account the simplifications and assumptions that have been made.

When we derived the rates of corrosion from the Evans diagrams, we assumed that the cathodic and anodic sites are equal in area. But if the corrosion is confined to particular regions on the metal surface (as in heterogeneous corrosion), then this situation no longer holds, and the areas of the anodic and cathodic sites can be very different. For example, if the area on which the cathodic reactions are taking place is 100 times greater than the area on which the anodic reactions are taking place, then, since $|I_{ca}| = |I_{an}|$ or $|A_{ca}i_{ca}| = |A_{an}i_{an}|$, the anodic current *density* must be 100 times greater than the cathodic current density. Thus, the rate of corrosion (related to the anodic current density), localized at one site, can be very high. This leads to the well-known *pitting* of metals, in which most of the metal surface remains intact, whereas corrosion at a single site can lead to perforation of the metal, which could render a component useless. The problem is discussed further in Section 4.2.

SAQ 4   By following similar arguments to those presented above, determine which cathodic reaction is likely to couple with the anodic reaction for the uniform corrosion of nickel at pH 0. Values of i_e (from Block 8, Table 1) are $10^{-5} \, \text{A m}^{-2}$ for the anodic nickel reaction (at pH 0) and $10^{-1} \, \text{A m}^{-2}$ for the cathodic hydrogen reaction. Obtain values of $E^\ominus$ from the S342 *Data Book*. Assume that: (a) the mechanism for the metal oxidation reaction is similar to the one proposed for iron; (b) values of i_e change with concentration in a similar way to those proposed for iron; (c) where data are not available, values are the same as for the corrosion of iron.

3.3   Passivation

So far, we have constructed Evans diagrams without reference to the corresponding Pourbaix diagrams. This is acceptable as long as the conditions fall *within* the corrosion domain of the Pourbaix diagram with a rather low possibility of the formation of a passive film. But if a passive film can be formed, then our analysis requires some modification.

When we built up an Evans diagram for the corrosion of iron at pH 6 (Figure 27), we concluded that the corrosion current density, if the iron anodic reaction couples with the oxygen cathodic reaction, is approximately $10^{-1.2} \, \text{A m}^{-2}$, and the corresponding corrosion potential is about $-0.54 \, \text{V}$ (relative to the S.H.E.). Examination of the Pourbaix corrosion diagram for iron (Figure 7), reveals that these conditions (pH 6 and $V_{corr} = -0.54 \, \text{V}$) do indeed lie in the corrosion domain for iron: our analysis cannot be complicated by passivation. But suppose we follow the Tafel line for the iron anodic reaction (at pH 6) towards more positive values of the potential difference: according to Figure 27, we would expect the current density to increase. Indeed, this situation should continue until, as the Pourbaix corrosion diagram (Figure 7) suggests, the potential difference reaches a value of approximately 0.05 V, at which point *passivation* starts to occur. (This value is obtained by drawing a vertical line at pH 6 on Figure 7 until the boundary between corrosion and passivation is reached. The lines intersect at a potential difference of 0.05 V.) Thus, at this value of the potential difference, Tafel behaviour is no longer maintained and the current density decreases substantially as solid oxide precipitates out, protecting the surface of the iron from further corrosion. The situation is depicted in Figure 28. We haven't met this sort of Tafel plot before!

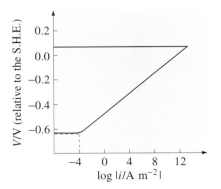

Figure 28   A Tafel plot for the iron anodic reaction at pH 6, showing the effect of passivation.

A more general diagram revealing this behaviour is shown in Figure 29, in which we have now plotted current density (rather than log |*i*|), in line with the usual practice for this sort of diagram. The actual experimental curve is smoother than would be expected from Figure 28. The loop labelled PQRS in Figure 29 is called the **active loop**. The value V_{pp} corresponds to the potential at which passivation occurs (at a particular value of pH): it is known as the **peak passivation potential**. The current density at this potential is called the **critical passivation current density**, i_{crit}. Notice that the current density doesn't drop to zero when the passive layer is formed. Between points S and T the current density has a constant value, which is the anodic current as metal continues to dissolve slowly through the passive film. This film is generally some 2–3 nm thick.

The size of the active loop is very sensitive to changes in the pH. Figure 30, which shows active loops for iron at various pH values, reveals that the loop decreases in size as the pH increases.

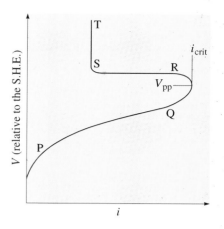

Figure 29 A general plot of *V* versus *i* for a metal undergoing passivation.

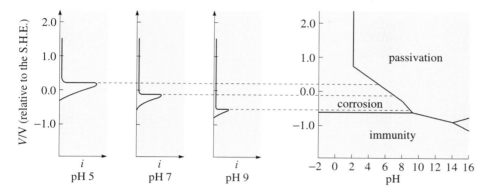

Figure 30 Active loops for iron at various pH values. Values for the peak passivation potential, at a particular value of pH, can be determined from the Pourbaix corrosion diagram for iron, which is also shown.

The size of the active loop (at a particular pH value) can also be changed by adding other elements to iron. For example, increased corrosion resistance is conferred by adding chromium and/or nickel to iron. The so-called **'stainless' steels** are alloy steels containing 13–20% chromium and up to 12% nickel. The Pourbaix corrosion diagram for a 12% chromium alloy is compared with that for pure iron in Figure 31.

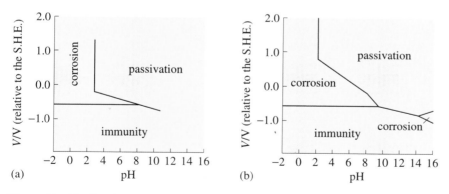

Figure 31 The Pourbaix corrosion diagram for a 12% chromium alloy (a), compared with that for pure iron (b).

The crucial point to note is the larger area of the passive domain in the chromium alloy. As you might expect, this affects the size of the active loop, as indicated by the comparison in Figure 32. As the percentage of chromium increases from zero to 20%, so the size of the loop *decreases* from alloy A1 (pure iron) to alloy A3 (20% Cr). If a curve for the oxygen cathodic reaction is superimposed, then an Evans diagram (of the first type discussed in Section 3.1, cf. Figure 19, but with current density plotted) is produced (Figure 33). Notice that the cathodic curve first meets the A1 and A2

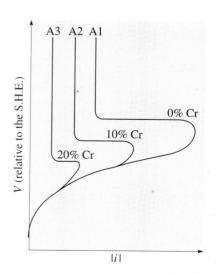

Figure 32 Active loops (at pH 7) for iron containing various percentages of chromium.

curves at point P. If the areas of the cathodic and anodic sites are equal, then this intersection point gives the value of the corrosion current density. Thus, under these conditions, corrosion without passivation occurs. This is the situation we have met previously; the curves cross where both reactions are following Tafel-type behaviour.

But concentrate now on the point of intersection for curve A3 in Figure 33. Here the curves intersect at a point (labelled Q) where the metal surface is undergoing passivation. As can be seen from the Figure, the corrosion current density (at pH 7) for alloy A3 is going to be very much less than the corrosion current density for the other alloys. Thus, under these conditions, the rate of corrosion of a 20% chromium alloy is very small, and much less than that of pure iron. It is observed that steels acquire this 'stainless' character in many common environments when the percentage of chromium exceeds 12%.

Thus, when calculating the corrosion current density from Evans diagrams, we must exercise caution when conditions are such that we are working close to the passivation domain in the Pourbaix diagram. For pure iron, these complications arise above about pH 8. But we must interpret this statement with care, for corrosion is a dynamic process: the concentrations change with time. Thus, even though the pH of the bulk of the solution may have a value of 8, the pH of the solution close to the iron could be higher, because *the pH increases during corrosion*. Further difficulties arise in the presence of chloride ions, which tend to break down the passive film once it has been formed. This topic is discussed further in Section 4.4.

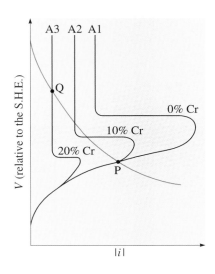

Figure 33   Evans diagrams for iron (at pH 7) containing various percentages of chromium, and showing the oxygen cathodic process.

3.4   Summary of Section 3

1   To explain corrosion processes, kinetic as well as thermodynamic data need to be considered.

2   During corrosion, the system behaves like a short-circuited electrochemical cell in which the potential differences across the two metal–solution interfaces are identical, and in which the net anodic current is equal to the net cathodic current.

3   By selecting appropriate sections of the i versus η curves for the net oxidation process, taking place on one part of the metal, and for the net reduction process, taking place on another part of the metal, and by separating these curves by the difference between their electrode potential values, an Evans diagram can be produced.

4   From the point of intersection of the two curves in an Evans diagram, values for the corrosion current and the corrosion potential (relative to the S.H.E.) can be determined. If the areas of the cathodic and anodic sites are the same, then the Evans diagram will provide a value of the corrosion current density.

5   Evans diagrams can be used to reveal the nature of the cathodic reaction, at various values of the pH. Calculations on corroding iron, for example, reveal that at pH 6 the main cathodic reaction is likely to be reduction of oxygen gas, whereas at pH 0, in unagitated solutions, the cathodic reaction is likely to be reduction of hydrogen ions. In agitated solutions, or under conditions in which the concentration of dissolved oxygen can be high, the cathodic reaction will be reduction of oxygen, even at pH 0.

6   If the area of the anodic site is very small compared with the area of the cathodic site, then the anodic current density can be very large compared with the cathodic current density. This situation leads to very rapid, intense corrosion at one site.

7   Evans diagrams are modified when the values of the corrosion potential and pH lie in the passivation rather than in the corrosion domain (as revealed by the Pourbaix corrosion diagram).

8   Projection of the Tafel line for an anodic reaction into the passivation domain produces a loop, known as the active loop.

9   For iron, and iron–chromium alloys, the size of the active loop decreases with increasing pH. For alloys containing more than 12% chromium under neutral (pH 7) conditions, the oxygen cathodic line cuts the active loop in the passivation region, indicating that these alloys will be fairly resistant to corrosion at pH 7.

the crossover point gives you the corrosion potential and current

4   TYPES OF CORROSION

4.1   Active-site corrosion

For corrosion to take place, all that is necessary is for a metal anodic reaction to couple with some cathodic reaction, such that the overall reaction is both thermo-dynamically and kinetically possible. The initial attack is usually at an active site. An active site in a pure metal tends to be a surface site where the metal lattice is imper-fect in some way, such as at a point defect. Corrosion also tends to start where a metal lattice has been deformed, as occurs when a piece of metal is cut from a larger sheet, or when a metal is bent or deformed. But it is impossible to predict the exact location of these active sites in a pure metal, or what their relative reactivities will be. Even two pieces of metal of the same size, shape and mass cut from the same sheet and immersed in the same solution, usually have different patterns of corrosion. Thus, the intensity of corrosion at one site on one piece of metal can be very different from that on a second, and apparently similar, piece of the same metal.

- ■ Does this mean that the corrosion current (or the corrosion current density) will be different for the two pieces of metal?

- ■ For two pieces of metal of identical surface area cut from the same sheet, the corrosion currents must be similar, so an *intense* attack limited to one region on one piece could be matched by *uniform corrosion* of lesser intensity on the other piece.

This brings out an interesting point. If corrosion attack is confined to one region of a sample, this effectively *protects* other regions of the metal from attack. If the region being attacked were to be drilled out or cut away then attack would be set up at new, previously unattacked sites.

Once corrosion has started at an active site, it can develop in a number of ways, depending on the nature of the corrosion products, the nature and concentration of the aqueous solution, the degree of oxygen availability, and on whether the metal sur-face is vertical or horizontal. The pattern of corrosion attack also changes if surface defects are introduced, either deliberately or accidentally. For example, scratching the surface of the metal will disturb the metal lattice and provide active sites for initial attack.

More dramatic effects can result if other elements, such as impurities or substances added during manufacture, are present in the metal, since these can affect both the pattern and the rate of corrosion. The presence of these elements necessarily creates deformities in the crystal lattice of the metal, and hence active sites. If the 'foreign' element is also more prone to oxidation than the metal, then a further enhancement in the activity of the site will result. Most commercially produced metals contain some impurities, the main ones present in iron usually being sulfur and phosphorus. The presence of sulfur in iron, even in concentrations as low as 0.1%, affects the corrosion behaviour dramatically. The sulfur, as well as being present in the elemen-tal form, also combines with the iron, and other elements present, to form a range of metal sulfides (MS). As particles of these sulfides become exposed during corrosion, they can react with water to form hydrogen sulfide ions, HS^-; a divalent metal sulfide, for example, is hydrolysed as follows:

$$MS(s) + H_2O(l) = M^{2+}(aq) + HS^-(aq) + OH^-(aq) \qquad (18)$$

These hydrogen sulfide ions adsorb on the surface of the iron, and form very active sites, thus catalysing the iron oxidation reaction and increasing the rate of corrosion. It is the lack of sulfur in the Delhi iron pillar (Figure 1) that is, at least partially, responsible for its remarkable state of preservation. But it cannot be this factor alone; otherwise we would expect wrought iron, which used to be made by a process that virtually excluded sulfur, to be relatively immune to corrosion in air. Regrettably this is not the case! Rather, the sulfur in the atmosphere, mainly present as SO_2, exerts a greater influence than any that may be present in the iron. It is the lack of sulfur in the atmosphere, together with the lack of moisture, that mainly accounts for the preservation of the Delhi pillar.

SAQ 5   Hydrogen sulfide ions increase the rate of corrosion of iron by a factor of 10 at pH 2 and at 25 °C. Given the following data and assumptions, determine graphically what change in the value of α or i_e, for the iron anodic reaction (equation 1), would bring about this increase.

Assume that:

(a)   the cathodic and anodic sites have equal areas;

(b) $IR_S = 0$;

(c)   the processes are not limited by concentration polarization or the formation of a passive film;

(d)   the cathodic reaction is unaffected by the presence of hydrogen sulfide ions;

(e)   at pH 2 the cathodic reaction is reduction of hydrogen ions, for which the transfer coefficient, $\alpha_{ca, red} = 0.5$;

(f)   the value of α for the oxidation of iron (in the absence of HS^-) is 1.5;

(g)   under corrosion conditions at pH 2:

for the cathodic reaction: $E(H^+|H_2) = -0.12\ V$; $i_e = 10^{-3}\ A\ m^{-2}$

for the anodic reaction: $E(Fe^{2+}|Fe) = -0.64\ V$; $i_e = 10^{-8}\ A\ m^{-2}$

[*Hint*   First of all, construct the Evans diagram (*V* versus $\log |i|$ type) for the anodic oxidation of iron and the cathodic reduction of hydrogen ions. Note the value of the corrosion current density. Increase this by a factor of 10 (by adding 1.0 to $\log i$) and mark this value on the line for the cathodic reduction of hydrogen ions. Now consider in what ways the line for the iron anodic reaction must change in order that it passes through this point.]

To convert iron into a useful structural material, the elements carbon, silicon and manganese are often added. The addition of carbon confers on the metal the possibility of heat treatment. Such heat-treatable iron alloys are known as steels. The so-called carbon steels are used for both structural and general engineering purposes. Additional alloying elements, principally chromium, nickel and molybdenum, are also sometimes added to give corrosion resistance (Section 3.3) and extra hardness.

4.2   Differential aeration corrosion

Whenever a metal is simultaneously exposed to two different moist environments, containing different amounts of oxygen, then it will undergo differential aeration corrosion in addition to active-site corrosion. The part relatively rich in oxygen will become the site of the cathodic reaction, whereas that relatively poor in oxygen will become the site of the anodic reaction. We met this sort of situation earlier on, when we considered the action of a drop of water on a horizontal sheet of iron (Section 1.2). After a time, the reactions settle down to a cathodic reaction occurring around the periphery of the drop, and the anodic reaction, oxidation of the metal, at the centre. To see how this phenomenon fits in with the previous Section, from which we would expect attack to occur preferentially at active sites, let's examine the process in more detail.

When a drop of water is placed on an iron sheet, corrosion — iron oxidation — *will start at active sites*, with the corresponding cathodic reaction taking place on adjacent sites. As reaction continues, however, oxygen gas dissolved in the drop is being used up. Oxygen will diffuse into the drop from the surrounding air, but areas around the perimeter of the drop will gain more oxygen than will areas toward the centre. As a result, the cathodic reaction will be starved of oxygen at the centre of the drop, and will eventually die away. At the same time, the more 'successful' cathodic reactions taking place at the edges consume oxygen and produce hydroxide ions via the equation:

$$\tfrac{1}{2}O_2(g) + H_2O(l) + 2e = 2OH^-(aq) \qquad (2)$$

Thus, the pH at the perimeter will increase, until it reaches a high enough value for iron(II) ions to be converted into solid oxides, according to the Pourbaix corrosion diagram for iron. Thus, any anodic reactions taking place at the edges of the drop will eventually be phased out as the sites become covered with a protective oxide layer. After a time, therefore, the anodic reactions will be confined to the centre of the drop, and the cathodic reactions to the periphery.

SAQ 6   Predict the final distribution of anodic and cathodic sites that will result from a drop of water being placed on a horizontal iron sheet if (a) the drop is saturated with oxygen before coming into contact with the metal; (b) the drop is depleted in oxygen before coming into contact with the metal; (c) oxygen is blown into the centre of the drop and nitrogen gas surrounds it.

The drop experiment is just one instance of corrosion by differential aeration. There are many others, some examples of which are discussed below.

4.2.1   Painted surfaces

As long as the surface is covered by paint, oxygen and hydrogen ion availability is limited, and corrosion is very slow. However, we all know the problem: painted surfaces are liable to be chipped or scratched! As soon as some metal is exposed to moist air, corrosion can take hold. The situation, as depicted in Figure 34, may not be as you envisaged it. The exposed metal has access to a higher concentration of oxygen than the metal surface *under* the paint. Thus, the cathodic reaction occurs at the exposed metal surface, and the anodic reaction occurs at the metal surface underneath the paint. This means that corrosion is taking place *underneath* the surface of the paint, so the situation may be more severe than superficial inspection reveals — and it will get worse! The rust formed eventually lifts off the layer of paint, and so the process can continue on pure iron under the paint surface around the edges of this bigger area of exposed metal, and so on.

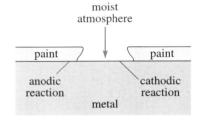

Figure 34   Metal exposed to the atmosphere through chipping of a painted surface.

4.2.2   Exposed surfaces with flaws

An untreated iron surface exposed to neutral aerated water will corrode, as we would expect from all our discussion so far. On such a surface, corroding by active-site corrosion, we would expect the rate of corrosion to be fairly uniform — usually less than 1 mm per year. As corrosion continues, oxygen gas is consumed via the cathodic reaction and hydroxide ions are produced. In time, the concentration of hydroxide ions on the surface of the metal will be sufficient to 'push' the iron into the passivation domain of the Pourbaix corrosion diagram, and the rate of corrosion will diminish. But if the iron contains a surface flaw or is damaged in some way, so that a *crevice* is formed on the surface, then the results can be disastrous. As shown in Figure 35, there will be limited access to oxygen within the crevice compared with the surface of the metal; thus, oxidation of iron will occur *within the crevice*. The crevice will get larger, there will be a further increase in the differential oxygen availability, and the process will continue.

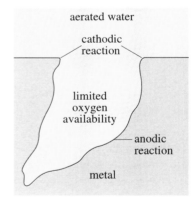

Figure 35   An untreated metal surface containing a crevice.

But the situation can well become even worse: the crevice, being the site of the anodic reaction, will attract anions. Thus, chloride ions, for example — ions that are abundant because of the salting of roads in the winter months — will migrate into the crevice. Here, any iron already in solution will be present as iron(II) because of limited oxygen availability. The initial product is an iron(II) chloride complex, $FeCl^+(aq)$, which hydrolyses to form the insoluble iron(II) oxide via the following reaction:

$$FeCl^+(aq) + H_2O(l) = FeO(s) + 2H^+(aq) + Cl^-(aq) \qquad (19)$$

The crucial point is that hydrogen ions are produced in this reaction. These will lower the pH of the solution in the crevice, further increasing the rate of corrosion. This sort of attack can be very severe and can lead to the destruction of components in a very short time, the bulk of the metal remaining virtually intact.

SAQ 7   Use the principle of differential aeration corrosion to predict the region of metal corrosion in the situations shown in Figures 36–38.

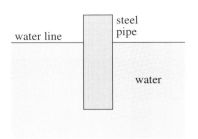

Figure 36   A steel pipe partially submerged in water.

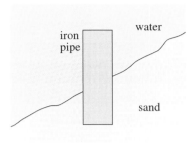

Figure 37   An iron pipe embedded below water in sand.

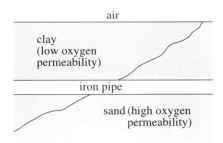

Figure 38   An iron pipe passing through soils of different oxygen permeability.

4.3   Galvanic corrosion

If two different metals are electrically connected and a cathodic reaction is possible, as shown in Figure 39, then the metal with the lower (more negative) electrode potential will preferentially corrode, with the cathodic reaction being confined to the other metal. Thus, if metal B has the lower value of electrode potential, then it will preferentially corrode (usually where it is electrically connected to the metal A), and the cathodic reaction will take place on metal A. This fact is actually made use of in cathodic protection, which will be discussed in Section 5.2. But, as a source of corrosion, it has been responsible for many catastrophic metal failures, and it is a common design error. The degree to which B is preferentially attacked depends not only on the difference between the electrode potentials of the two metals, but also on the relative areas of A and B. If the cathodic reaction is confined to A, and this has a very large surface area compared with B, then (since $|A_{ca}i_{ca}| = |A_{an}i_{an}|$) the anodic current density can be very large, leading to rapid corrosion. This has been observed when steel junctions have been used in copper tubing, and when aluminium rivets have been in contact with steel panels.

Which of the two metals will be preferentially attacked depends on their relative electrode potentials. However, standard electrode potentials provide only a rough guide, because they relate to pure metals in solutions of their ions (at unit activity). A more practical table, which takes account of commercial alloys in real environments, is shown in Table 1 — in this case for alloys immersed in seawater. The table is known as a **galvanic series**[*]. The alloys are grouped according to the extent to which they enhance corrosion of metals lower in the series. Notice also that several alloys, stainless steel for example, are listed twice — once in a passive form (when the metal is covered with a passive film), and again, in an active form (when the passive film has not been allowed to form or when it has been destroyed in some way). It can be assumed that galvanic effects are nil for pairs of alloys in the same group in the series, but increasingly significant with increasing separation within the series.

An example of galvanic corrosion is the damage that occurred on a large oil tanker in the 1970s. The engine-cooling water was seawater, admitted by 'butterfly' valves (similar to the air control valve in a carburettor) manufactured from gunmetal. Gunmetal is a copper alloy, containing 10% tin: it has excellent corrosion resistance in seawater. Various rubber seals were secured to the gunmetal by bolts, screwed into tapped holes. In the larger valves, these bolts were made of stainless steel and no corrosion was observed. However, in the smaller valves, the bolts were made of carbon steel. The carbon steel and the gunmetal together constituted an excellent bimetal couple (see Table 1); the steel–gunmetal area ratio was about 1 : 50. Severe corrosion took place during only a few months service! As a result, the carbon-steel bolts had to be replaced by new ones made of stainless steel.

[*] Named after Luigi Galvani, one of the pioneers of electrochemistry at the end of the eighteenth century.

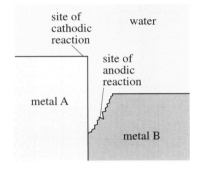

Figure 39   Galvanic corrosion.

Table 1   Galvanic series for alloys immersed in seawater. Alloys lower down in the series are preferentially attacked when coupled to alloys higher in the series.

titanium alloys (passive)

nickel alloys (passive)

stainless steels (passive)

silver alloys

copper alloys

lead–tin alloys (solders)

nickel alloys (active)

stainless steels (active)

cast irons

carbon steels

cadmium

zinc alloys

aluminium alloys

magnesium alloys

SAQ 8 Estimate the severity of galvanic corrosion for the following bimetal couples, indicating in each case which metal is attacked: (a) copper and titanium; (b) copper and solder; (c) silver and cadmium; (d) magnesium and aluminium; (e) cast iron and zinc; (f) stainless steel and magnesium.

SAQ 9 Domestic central heating systems often have copper tubing connected to cast iron or steel radiators. Suggest reasons why the 'moderate' corrosion expected from the galvanic series does not usually occur in such systems.

SAQ 9 reveals the influence of the relative areas of anodic and cathodic sites, and, more importantly, the closed nature of the water-circulation system. In large, industrial or commercial heating systems, the volume reduction in the water that occurs whenever the system is allowed to cool down (for example, in large office blocks at the weekend) can result in considerable replenishment of fresh water containing dissolved air. This situation is exacerbated when the heat-exchanger cylinder is made of zinc-coated (galvanized) steel rather than copper. In these circumstances, perforation of the cylinder can occur in a matter of months, a problem that can affect galvanized steel storage tanks in domestic water systems.

Much depends on the character of the water. Hard waters tend to deposit a protective chalky scale, which protects the iron or zinc components. Very soft waters, however, cannot do this, and, if slightly acid, have the property of slowly dissolving copper unless carefully treated. Such water supplies can galvanically deposit minute crystals of copper by a simple displacement reaction:

$$Zn(s) + Cu^{2+}(aq) = Zn^{2+}(aq) + Cu(s) \qquad (20)$$

$$Fe(s) + Cu^{2+}(aq) = Fe^{2+}(aq) + Cu(s) \qquad (21)$$

These crystals set up a host of tiny bimetallic zinc–copper or iron–copper corrosion cells wherever air is available as a cathodic reactant. The degree of surface coverage by these copper crystals can be substantial, and in some cases the area of exposed zinc (or iron) can become much less than the area of exposed copper. Bimetallic corrosion of zinc (or iron) can then occur, to the extent that zinc-coated steel tanks in areas where the water supply is acidic have been known to fail within weeks of being installed!

4.4 Pitting corrosion

The conditions favouring the passivation domain for various metals are revealed by their respective Pourbaix corrosion diagrams. However, the existence of a passivation domain does not *guarantee* inert behaviour under these conditions. The passive film can break down at small areas on the surface, leading to localized corrosion that is very destructive. This **pitting corrosion** penetrates inwards more rapidly than it spreads sideways (compare crevice corrosion, Figure 35) so that the serviceability of a component can be destroyed as a result of a single pit penetrating its entire thickness, whereas the rest of the metal appears unaffected. The reason for this breakdown of the passive film is almost invariably the presence of chloride ions. Chloride ions apparently have the ability to adsorb on the passive film, and to exchange with anions in the film — replacing oxide ions, for example. This is thought to damage the regularity of the oxide lattice, and so allow metal ions to diffuse through the lattice, thereby increasing the rate of corrosion.

The effect can be seen from an examination of the V versus i plot. For a metal entering the passivation domain, we would expect the V versus i plot to look something like Figure 29, repeated here as Figure 40a. However, the presence of chloride ions affects the current density *once the passive layer has been formed*. This is illustrated schematically in Figure 40b, which shows the effect on the curve as the chloride ion concentration is increased (curve A1 corresponds to the behaviour in the absence of Cl⁻). The points labelled X, Y and Z are called the **breakdown (or pitting) potentials**, and are given the symbol V_b. For any particular chloride ion concentration, any potential more positive than V_b will result in breakdown of the passive film, and an increase in the corrosion rate. The passive film will be destroyed

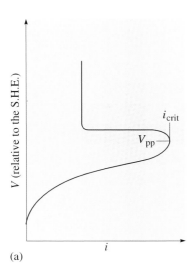 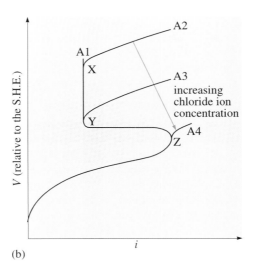

Figure 40 The effect of increasing chloride ion concentration on the plot of *V* versus *i* for a metal undergoing passivation: (a) in the absence of chloride ion; (b) with various chloride ion concentrations.

at active sites on the surface, leading to localized pitting attack. For this reason, V_b is also known as the **pitting potential**.

Other anions can have a similar effect to chloride ions: the most notable ones are sulfate ions, which attack carbon steel, and fluoride ions, which attack alloy steels.

An example of pitting attack was found in a stainless steel refrigerator tube in a large storage plant. The pitting, which was unexpected, was confined to the bends in the tubing, and refrigerant had leaked in several places, causing severe contamination of the store contents. Prior to bending, the tubes had been filled with low-melting resin; afterwards, the resin was melted out and the tube interior was degreased by hanging it in trichloroethene vapour; finally, all the bits were welded together. The interior of the pits was found to contain chloride: but where had it come from? There was no chloride in the original resin but there *was* some in traces of the resin left in the tube bends. The most probable source was the trichloroethene. Vapour degreasing is a well-established technique, but if the solvent is heated over several months without replenishment, it slowly decomposes, yielding HCl. When the residual particles of resin were exposed to the degreasing vapour, the HCl present reacted with them to form small 'pockets' of chloride in contact with the tube interior. All that was needed to complete the tale of woe was wet nitrogen gas, used to flush the system before it was filled with refrigerant; this provided an aqueous chloride environment wherever traces of resin remained.

4.5   Weld corrosion

Stainless steels containing the appropriate percentage of chromium, for example, are corrosion resistant under most conditions. However, where stainless steel is *welded* to other pieces of *identical* stainless steel, corrosion can take place along the welded joints, leading to the pieces of metal falling apart. The reason for this seems to be as follows. During the welding process, in which molten metal is fed into the gap between the two items to be joined, the surrounding area becomes very hot. The high temperature allows chromium to react with carbon in the steel to form chromium carbide at the boundaries (Figure 41). The regions adjacent to these boundaries become depleted in chromium, and are therefore more susceptible to attack.

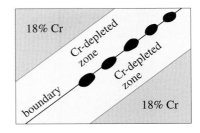

Figure 41 An example of weld corrosion. The black areas represent chromium carbide.

4.6   Erosion corrosion

The rate of corrosion depends, among other things, on the rate of the cathodic reaction. If this is the reduction of oxygen dissolved in the aqueous phase, then the rate of corrosion will, as we've seen, depend on the availability of oxygen. In unagitated, neutral, aerated water, the rate of corrosion of iron is limited to about 0.5 mm per year. The rate of corrosion will increase if the aqueous medium is stirred in some way or — a more common situation — if it flows through, or over, the metal surface. This is all to be expected. However, if the flow rate is much above 5 m s^{-1}, the smooth liquid flow can give way to turbulence. When this happens, the static layer of liquid adjacent to the metal surface is grossly disturbed, and this can lead to the passive film being torn away from the metal surface. Rapid corrosion then ensues, the rate being further enhanced by the rapid flow of solution, which carries oxygen to the adjacent cathodic sites. The effect of so-called **erosion corrosion** is similar to pitting corrosion, except that the damage to the passive film is now mechanical rather than chemical.

Typical examples of erosion corrosion are the attack on valves, pumps and heat exchangers experienced in a cooling plant if insufficient care has been taken to limit turbulence. One example that can be cited is the outlet pipe from a liquid fertilizer tank containing concentrated ammonium nitrate solution, which normally passivates mild steel (oxidizing it to deposit a film of iron(III) oxide). Here, however, there was a sharp bend in the pipe, which gave rise to extremely turbulent conditions, even though the flow elsewhere was relatively smooth. The pipe lasted exactly two months — a corrosion rate of 40 mm per year!

4.7   Bacterial corrosion

Metals buried in soils can corrode both via active-site corrosion and as a result of differential aeration. In acidic soils, hydrogen ion reduction is a possible alternative cathodic reaction to oxygen reduction, as we might expect. But in many soils, bacteria provide a further alternative cathodic reaction, so that even in neutral anaerobic soils (such as those at the bottom of the North Sea, for example) corrosion is still possible. The bacteria of particular interest are *Desulfovibrio* — or in English, *sulfate reducers*. These bacteria are able to use sulfate ions present in the soils to generate a cathodic reaction:

$$SO_4^{2-}(aq, soil) + 8H^+(aq) + 8e(bacterium) = S^{2-}(aq) + 4H_2O(l) \qquad (22)$$

This cathodic reaction can couple with the iron anodic reaction to produce iron(II) ions; that is, the iron corrodes. The iron(II) ions combine with the sulfide ions to give a black precipitate of iron(II) sulfide. Thus, iron attacked in this way appears black, rather than the more usual rust colour. The bacteria contain enzymes that speed up the reaction, with the result that, in polluted soils and sea silts, iron corrodes unexpectedly quickly. As you might expect, the reactions involved are actually a good deal more complicated than is indicated above.

4.8   Stray-current corrosion

Iron buried in soil normally rusts at a rate of something between 0.02 and 0.2 mm per year. Differential aeration, with a small anodic/cathodic ratio, can push this up by a factor of five, and bacterial corrosion can also have an effect under some circumstances. Occasionally, however, even higher rates of corrosion are observed — when so-called 'stray currents' are present in the soil. These are electrical currents that 'leak' through the soil, either from traction systems (electric railways, tramways, etc.) or from neighbouring buried structures that are being protected by cathodic protection (see Section 5.2). If a buried iron structure, such as a pipeline, provides a path of low resistance, it conducts the stray current, which may set up a potential difference between two parts of the iron. A cathodic and an anodic area then develop, and very rapid corrosion can take place at the anodic area. Avoiding such problems requires close cooperation between the various interests involved, such as workers in industries like electricity, gas, water, telephone and transport. In fact, a national body, the Joint Committee for the Coordination of the Cathodic Protection of Buried Structures for Great Britain, was formed in 1953 to provide such cooperation.

The above discussion reveals that there is a large number of different types of corrosion. We have considered one or two examples relevant to each particular type but, of course, corrosion cannot always be easily categorized in this way. Any corroding system could be subject to all eight ways mentioned, so it may seem surprising that any metal lasts for any significant length of time! However, just as many techniques have been developed to combat corrosion. These are discussed in the next Section.

5 CONTROLLING CORROSION

5.1 Design

By this stage, you should have an inkling that, although corrosion is a quite simple electrochemical process, it is often governed by complicating factors, some of which are neither chemical nor electrochemical. For instance, component geometry is important in determining the access of oxygen, and rates of attack are thereafter controlled by diffusion of oxygen. Other important factors include: the chemical composition, electrical conductivity and rate of flow of the electrolyte; the ease, or otherwise, with which surface films are generated, their mechanical strength and porosity; the chemical composition, homogeneity and microstructure of the metal, and the state of its internal stress. Corrosion, therefore, is a cross-disciplinary subject, embracing various branches of physical chemistry, metallurgy and engineering.

In the light of the examples in Section 4, it will come as no surprise to discover that there is no Magic Lozenge for stopping corrosion. Instead, many small things can be done to keep corrosion within the bounds of economic acceptability. The most obvious is to avoid complicating factors, such as adverse geometry, or erosion, or mixed metals. By corrosion-conscious design, the engineer can almost always avoid crevices, excessively rapid electrolyte flow and galvanic couples. If careful attention is paid to the selection of materials, the minimizing of unnecessary mechanical stresses, and the provision of drainage holes, then corrosion does in fact become simple: it becomes *uniform* and, with a suitable choice of alloys, slow. It is then that the Evans diagram and electron-transfer kinetics can be applied to predict rates with reasonable reliability. Nevertheless, as we shall see below, other measures can also be adopted to extend a component's working life. Thereafter, it is largely a matter of trying to minimize human error (such as failing to use a specified material or follow a specified procedure), or of changing the operating conditions. The video sequence associated with this Topic Study (band 9 on videocassette 2) shows many of these methods of corrosion control being used in the manufacture of motor cars.

5.2 Cathodic and anodic protection

You will remember that any given metal/metal ion couple displays an equilibrium potential given by the Nernst equation. Below that potential, oxidation of the metal to the ion in question is thermodynamically impossible, and we reflect this fact in the Pourbaix corrosion diagram (Figure 7, for example) by labelling this region 'immunity'. As stated earlier (Section 2), an effective method of control is referred to as **cathodic protection**: it can be achieved in one of two ways.

The first exploits the phenomenon of galvanic corrosion, by connecting the system to be protected to a more-reactive metal, thereby obtaining **'sacrificial' protection**. For example, to protect large steel structures such as harbour installations, oil production platforms, ships and underwater pipelines, it is common practice to connect lumps of zinc suitably spaced along the structure. The connection must of course be metallic, to permit the flow of electrons, as indicated in Figure 42. The zinc, appearing below iron in the galvanic series (Table 1), corrodes preferentially, at a lower (that is, more-negative) value of the corrosion potential. If the corrosion potential, V_{corr}, is more negative than $E(Fe^{2+}|Fe)$, as indicated in Figure 43, then iron corrosion is thermo-dynamically impossible. In view of the finite electrical resistance of the electrolyte, the local potential of the steel rises with increasing distance from the zinc, and this determines the spacing of the sacrificial metal.

An alternative means of achieving the same result is to coat the steel structure with a thin layer of zinc. Such **galvanized coatings** are used chiefly under conditions of atmospheric exposure, where the rate of zinc corrosion is sufficiently low to give protection over a period of several years (compare Section 5.5).

The second method of cathodically protecting metals is via an *impressed current*. Here, the potential is lowered by supplying an external source of direct current. The

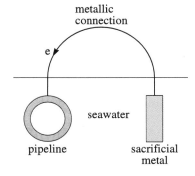

Figure 42 Cathodic protection using a sacrificial metal.

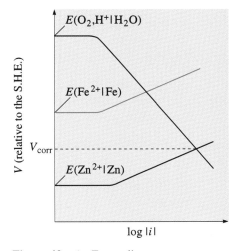

Figure 43 An Evans diagram showing the use of zinc as a sacrificial metal. Zinc has a more-negative value of V_{corr} than iron.

metal to be protected is connected to the negative pole, and the 'inert electrode' is connected to the positive pole (Figure 44). Like the sacrificial-metal technique, it is usually necessary to have several inert anodes, each with its own source of direct current, strung out at intervals along the structure. Naturally, it is of some practical importance how much current is needed to effect protection, just as it is important to know how long the zinc will last before it requires replacement in a sacrificial-metal system. To minimize the current requirement, it is customary to employ inert surface coatings in combination with cathodic protection. In this way, the cathodic protection system is needed to protect only bare areas of steel exposed at defects in the coating.

Although cathodic protection might seem an ideal way of preventing corrosion, it does have some drawbacks. Firstly, when external current sources are used, the power consumption necessary to achieve protection may be impractically large. Secondly, the current may not be uniformly distributed over the corroding metal, in which case there could be local areas at which the potential difference is not lowered enough to take the metal into the immunity domain. Thus, localized corrosion could be taking place on a metal that is thought to be fully protected. Another problem can arise if the metal is made so excessively negative that the rate of the alternative cathodic process, hydrogen ion reduction, becomes appreciable. The hydrogen gas formed on the metal surface may dissolve into the metal, reducing its internal strength and making it more brittle.

Anodic protection is different in principle. Its purpose is to *raise* the potential of a metal into the *passivation* domain of the Pourbaix corrosion diagram. Accordingly, the system being protected is connected to the *positive* pole of a source of direct current, and the 'counter electrode' is an inert metal, serving as a cathode (Figure 45). We saw earlier (Figure 29, Section 3.3) that a current density of at least i_{crit} is needed to effect passivation, so it follows that for an extensive structure, very large currents would be necessary to produce the passive condition. Moreover, such a technique could succeed only where the current is very small once the passive film is formed. This method is, therefore, used only when the metal is already passivated, and so is confined largely to titanium alloys and stainless steels in chemical plant: the anodic protection serves to strengthen the passive film and repair it when necessary. One apparent exception is the anodic protection of mild steel in 'oleum', the SO_3-saturated sulfuric acid initially produced in the contact process. Curiously, ordinary iron passivates easily under these conditions.

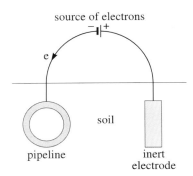

Figure 44 Cathodic protection using an impressed current.

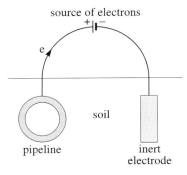

Figure 45 Anodic protection using an impressed current.

5.3   Water treatment

One way to ameliorate corrosion in closed (for example, recirculatory) systems is to treat the water itself with soluble **inhibitors**. These are of two main kinds. The first is a reagent that removes oxygen. In Section 4.6 we saw that the rate of corrosion in unagitated, neutral aerated water is limited (to about 0.5 mm per year) by the rate of diffusion of dissolved oxygen. Since the maximum rate of diffusion of any species is proportional to its concentration, it follows that a 500-fold reduction in dissolved oxygen content should lower the rate of corrosion to 10^{-3} mm per year, while at the same time preventing differential aeration attack. Some of the dissolved air can be removed simply by exposing the water to a partial vacuum. However, virtually total removal follows if suitable reducing agents are then added, such as sulfite ions or hydrazine. Sulfite, for example, reacts with dissolved oxygen to form sulfate, via the following reaction:

$$2SO_3^{2-}(aq) + O_2(g) = 2SO_4^{2-}(aq) \tag{23}$$

The second kind of inhibitor is one that promotes passivity. A glance at the corrosion diagram for iron (Figure 7, Section 2) should remind you that passivity is encouraged by raising either the pH (but not too much!) or the potential difference. Passivating inhibitors are therefore of two main kinds: those that are hydrolysed to give hydroxide ions (raising the pH), and those that provide an alternative cathodic reaction, which raises the value of V_{corr}. Apart from NaOH itself, those in the first category are the sodium salts of weak acids, such as Na_2CO_3 and Na_2SiO_3. They possess the further advantage of yielding anions that are themselves passive film promoters. The salts are first hydrolysed, which raises the pH of the solution; for example

$$Na_2SiO_3(aq) + 2H_2O(l) = H_2SiO_3(aq) + 2Na^+(aq) + 2OH^-(aq) \qquad (24)$$

The ions produced then react with the product of the anodic reaction, such as Fe^{2+} ions:

$$Fe^{2+}(aq) + SiO_3^{2-}(aq) = FeSiO_3(s) \qquad (25)$$

The resulting solid forms a protective coating on the metal surface.

Passivating inhibitors in the second category act in much the same way as anodic protectors; that is, they raise the corrosion potential (from V_{corr} to V_{corr}', say) into the passivation domain, as indicated for the nitrite ion (equation 26) in Figure 46:

$$NO_2^-(aq) + 8H^+(aq) + 6e = NH_4^+(aq) + 2H_2O(l); \qquad E^\ominus = +0.89 \text{ V} \qquad (26)$$

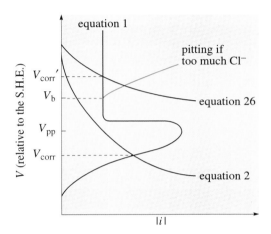

Figure 46 An Evans diagram for iron corrosion, showing the effect of changing the oxygen cathodic reaction (equation 2) to a nitrite cathodic reaction (equation 26).

Another commonly used ion is hydrogen chromate, which reacts as follows:

$$HCrO_4^-(aq) + 7H^+(aq) + 3e = Cr^{3+}(aq) + 4H_2O(l); \qquad E^\ominus = +1.38 \text{ V} \qquad (27)$$

However, it is important not to forget the problems that result if 'aggressive' anions, such as chloride, are present in the solution. As you saw in Section 4.4, the higher the chloride concentration, the lower is the breakdown potential V_b (recall Figure 40b). It follows that the balance between oxidizing anions and chloride is critical: too much chloride in the water, and the addition of inhibitor serves only to *promote* pitting corrosion, by raising the corrosion potential above V_b to V_{corr}', as indicated by the upper broken line in Figure 46.

To overcome this problem of reliability, it is common practice to use two or more inhibitors in combination. For example, for carbon steels, some 20×10^{-3} mol dm^{-3} Na_2CrO_4 may be used with 5×10^{-3} mol dm^{-3} Na_2SiO_3 or Na_2HPO_4, in waters containing up to 5×10^{-3} mol dm^{-3} chloride ion. Larger concentrations of chloride require larger additions of inhibitor, and there finally comes a point at which reliability can be restored only by the use of chloride-resisting alloys, such as certain grades of stainless steel or copper alloys. Very high concentrations of additives can interfere with heat transfer, and in any case it is difficult to achieve uniform concentrations throughout the system.

5.4   Chemical conversion coatings

'Passivation' may be described as the conversion of a metal surface into a relatively inert compound. If this compound is in the form of a thin continuous film, it constitutes a high electrical resistance in the corrosion cell: the way in which this reduces the corrosion current should be clear from Figure 23 (Section 3.1). Passive films are normally oxides or hydroxides, and their thickness is in the region of 1 to 5 nm. However, the composition of the passive film depends critically on the environment and, if the electrical potential gradients are sufficiently large, the film can achieve much greater thicknesses of the order of 1 to 10 μm. Passive films may be produced either by controlled corrosion reactions, or else by subjecting the metal to electrolysis.

Typical conversion coatings are phosphate for steels, chromate/oxide for aluminium, and fluoride for magnesium. The metal is given an oxidizing chemical treatment to produce a layer that resists further attack. For example, **phosphating** of iron, cadmium plate or aluminium is effected by placing the component in a bath containing phosphoric acid and either zinc phosphate or manganese phosphate, which results in the formation of a complex surface phosphate of iron and zinc (or manganese). To allow the process to occur reasonably quickly (15 minutes) at room temperature, oxidizing agents such as nitrate may be added, without apparently harming the film so formed.

The electrolysis of aluminium sheets in sulfuric acid solutions is known as **anodizing** because the aluminium serves as the anode. A film of aluminium oxide is produced, whose conductivity is much less than that of the metal. The aluminium oxide film can grow to thicknesses of up 30 μm if the potential gradient is large. The structure of the film is complex. There is an inner, 'barrier' layer, akin to a passive film. Outside this there is a porous layer, which needs to be sealed by a hydrolysis treatment before the coated metal is put into service. Anodized aluminium coatings can withstand many years' exposure to mild atmospheres, but they fail by pitting after only a few months in industrial or marine atmospheres.

5.5 Alloying and metal coatings

The ideal metal should be strong, tough, cheap and inert. Unfortunately, the more-inert metals are among the most expensive, and iron, the cheapest and most abundant, is disappointingly reactive. All is not lost, however. To achieve durability, a useful compromise can be reached by modifying the surface layers of iron so that it either contains atoms of an inert metal or else is coated with a layer of the same. The first technique can most readily be achieved by **alloying**. The second involves **metal coatings**, applied in a variety of ways. Underlying all these techniques is the idea that the essential strength, electrical properties and so on are provided by the metal or alloy, and that, because corrosion is a surface reaction, all that is required is some form of protection *in situ*.

It might be thought that the ideal alloying elements for corrosion resistance would be those that are most inert, such as the platinum-group metals, or perhaps gold or silver. This proves not to be the case, chiefly because of an atomic-scale mechanism of selective dissolution resembling galvanic corrosion. In other words, the iron–platinum mixture (say) constitutes a bimetallic couple, so that iron continually leaches out through a porous surface layer of platinum. What does work, however, is the addition of *alloying elements that promote passivity*. For example, chromium promotes the passivation domain of iron, and progressively reduces the active loop as its concentration is increased (Figure 32, Section 3.3). For the same reason, additions of tin to copper (to form bronze or gunmetal) markedly improve its corrosion behaviour. Often, the degree of passivity depends critically on quite minor additions. Of the various grades of stainless steel, that containing about 3% molybdenum is the most resistant to pitting in chloride solutions. Similarly, the addition of about 1% each of iron and manganese to nickel brass (copper alloyed with 10–20% nickel) confers unexpectedly high resistance to corrosion by estuarial waters. The precise mechanisms are not always understood: what is clear is that these minor alloying additions stabilize the passive film against the undermining action of polluting anions.

In considering metallic surface coatings, we shall examine three typical coatings for iron, which are normally applied in three different ways. The first is **aluminizing**, in which the steel article is heated at 950 °C in a pack containing alumina (Al_2O_3) particles, molten aluminium and a chloride catalyst. A major role for the chloride is to form gaseous aluminium chloride. This has a covalent, dimeric structure, and acts as a transfer intermediate by decomposing on the iron surface:

$$Al_2Cl_6(g) = 2Al(Fe) + 3Cl_2(g) \tag{28}$$

The aluminium diffuses into the iron to give an aluminium-rich outer layer, which behaves to all intents and purposes like aluminium itself. It therefore passivates readily, and withstands atmospheric exposure at temperatures up to 1 000 °C.

The second example is **galvanizing** (Section 5.2), in which the steel article is 'hot dipped' in liquid zinc at just below 500 °C. Again, a flux is needed to maintain metal/metal contact, and, again, the liquid metal reacts with the iron and diffuses inwards to produce a complex zinc-rich surface layer some 50 μm thick. The high reactivity of zinc might lead one to suppose that such coatings would have a very short life. However, in polluted atmospheres, zinc slowly acquires a passive film of chloride or sulfate. Failure occurs chiefly as a result of attack by acid rain, or because of differential aeration effects.

Both aluminium and zinc coatings, although they confer protection because of their passivating properties, owe a major part of their beneficial effect to galvanic protection. Thus, if for any reason the underlying steel is exposed (at the edge, for example, when the galvanized sheet is cut), the fresh zinc surface corrodes preferentially, and so protects the steel (Figure 43, Section 5.2).

The third example is a two-layer electroplated coating of nickel–chromium, misleadingly called 'chrome plate'. The details vary, but, typically, such coatings comprise a 20 μm layer of nickel, over which is plated 1–5 μm of chromium. As you can see from the galvanic series in Table 1, nickel is protected from corrosion relative to carbon steels. If exposed to the atmosphere, however, nickel develops a pale green 'tarnish' film of nickel salts so, for a better decorative effect, the further application of a chromium layer provides a bright, corrosion-resistant outer coating. The chromium, of course, passivates and so behaves like passive titanium or passive stainless steel in Table 1. Unfortunately, as normally practised, chromium plating is a process leading to a layer that is extensively cracked, which exposes the (relatively) reactive nickel. This constitutes a galvanic cell having an adverse anodic/cathodic area ratio, so that, unless steps are taken to prevent it, the nickel is soon penetrated. Those steps usually consist of (a) increasing the anodic/cathodic ratio, by deliberately putting additional cracks into the chromium, or (b) by reducing the inwards penetration of the nickel, by making it extra reactive, so that the attack spreads sideways. This last step involves incorporating sulfur into the final 1–2 μm of nickel during manufacture: the increased reactivity is akin to that proposed for sulfur in iron (Section 4.1). When, finally, the nickel layer *is* penetrated, the underlying steel is much more reactive, and so the presence of the coating actually *enhances* corrosion. This demonstrates the dangers inherent in using relatively inert coatings. To be truly successful, such coatings must be entirely free from defects.

5.6   Paints

Paints are, perhaps, the most varied and complex of all anticorrosion coatings. In essence they consist of an organic polymer film (the 'binder'), containing a filler ('pigment') dispersed throughout its thickness. In order to ensure freedom from through-thickness pores, it is customary to apply several layers, by brush, roller or spray, taking care to ensure that successive layers bond to one another to give an integrated coating. Most of the organic binders are initially dissolved in a solvent that prevents the polymerization process taking place while the paint is being stored, facilitates spreading, and then evaporates as the paint 'dries'. The degree of polymerization is governed by the frequency of cross-linking in the polymer: the more extensive the cross-linking, the stronger and more stable the final film (and, consequently, the more expensive). Some solvent-free paints, which contain epoxides, are catalysed just prior to application and 'dry' (that is, polymerize) at a rate depending on the temperature and catalyst concentration.

'High-build' paint coatings, some 300 μm per layer, can be designed by combining epoxide resin with long-chain organic fillers such as pitch. However, most pigments are inorganic, and the coatings are only about 100 μm per layer when dry. The pigments used are of three kinds: inert, inhibitive and sacrificial. **Inert pigments**, as the name implies, serve no other function than to thicken the paint film and so lengthen the ionic diffusion path. Micaceous iron oxide (MIO) is typical of this class, the flat platelets of pigments lining up parallel to the metal surface as the solvent evaporates. **Inhibitive pigments**, for example Pb_3O_4 ('red lead'), $ZnSiO_3$ or $ZnCrO_4$, act as a controlled slow-release source of inhibitive anions (compare Section 5.3) and are therefore incorporated in the 'primer' coat in contact with the metal. The zinc cation is thought to control the degree of hydrolysis by water that penetrates the film, and may also help to deactivate sulfide ions originating from the steel (Section 4.1) by precipitating zinc sulfide. Metallic zinc, in the form of a dust, is incorporated into 'zinc-rich' primer paints, an example of the use of a **sacrificial pigment**. The other principal sacrificial metal is aluminium. Both require the pigment to be present in sufficient proportions to ensure continuous metal contact throughout the coating thickness.

It is a common observation that a paint coating is only as effective as the degree of surface preparation allows it to be. This is partly a matter of adhesion, which is destroyed by residual grease or 'millscale' (the iron oxide scale that forms on steel when it is shaped at white heat in a rolling mill). A more insidious agent is rust on steel. Rust itself is not particularly deleterious if mechanically scrubbed to remove loose particles. The main damage is caused by the sulfate and chloride anions that it harbours, more especially in surface pits. If they are not removed, these anions remain, leading to contamination of the pigment and prevention of passivation. Hence, it is very important to remove them before painting begins. 'Blast' cleaning, which employs hardened steel or slag grit entrained in a stream of high-pressure air or water, is an effective method of doing this, while at the same time roughening the surface to improve mechanical 'keying' of paint and metal.

Normal practice, therefore, is first to prepare the metal surface, removing grease, loose rust and contaminating salts. Acid etching or blast cleaning may be undertaken to give an additional mechanical bond to the subsequent paint coating; sometimes the metal may be phosphated or chromated (Section 5.4) for the same purpose. An active primer coating is then applied (Figure 47). The rest of the coating is there *to protect the primer coating*, mainly by reducing the degrading effects of ultraviolet radiation, rain or abrasion.

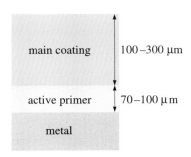

Figure 47   A typical paint coating.

OBJECTIVES FOR TOPIC STUDY 3

Now that you have completed Topic Study 3, you should be able to do the following things:

1   Recognize valid definitions of, and use in a correct context, the terms, concepts and principles printed in bold in the text and collected in the following Table.

List of scientific terms, concepts and principles used in Topic Study 3

Term	Page no.
active loop	24
active-site corrosion	26
alloying	36
aluminizing	36
anodic protection	34
anodizing	36
bacterial corrosion	32
breakdown (or pitting) potential, V_b	30
cathodic protection	33
chemical conversion coatings	35
corrosion current, I_{corr}	15
corrosion current density, i_{corr}	17
corrosion potential difference, $\Delta\phi_{corr}$	15
critical passivation current density, i_{crit}	24
differential aeration corrosion	7, 27
erosion corrosion	31
Evans diagram	17
galvanic corrosion	29
galvanic series	29
galvanized coatings	33, 37
heterogeneous corrosion	7
homogeneous corrosion	7
immunity	10
inert pigments	37
inhibitive pigments	38
inhibitor	34
metal coatings	36
paints	37
passivation	10, 23
peak passivation potential, V_{pp}	24
phosphating	36
pitting corrosion	30
Pourbaix corrosion diagram	10
Pourbaix diagram	9
sacrificial pigments	38
'sacrificial' protection	33
stainless steel	24
stray-current corrosion	32
weld corrosion	31

2   Given appropriate information, construct and interpret Pourbaix and Evans diagrams. (SAQs 1–5)

3   Describe the various types of corrosion, and, given a particular situation, determine the likely type and site of corrosion. (SAQs 6–9)

4   Describe the various methods used to reduce the rate of corrosion. (band 9 on videocassette 2)

SAQ ANSWERS AND COMMENTS

SAQ I (Objective 2)

We know from our everyday experience that gold is highly resistant to corrosion and remains untarnished under normal conditions. Thus, we would expect the immunity domain for gold to be extensive, which restricts the choice of diagrams to Figure 12c, and possibly Figure 12a. As rhodium is *less* susceptible to corrosion than gold, one of these two diagrams must be the Pourbaix corrosion diagram for rhodium. Suppose that in Figure 12a the upper domain represented corrosion. Then, the metal concerned would corrode when in contact with oxygenated water at any pH, something which does not happen with gold (nor, therefore with rhodium), so this must be a passivation domain. As gold is more susceptible to corrosion than rhodium, Figure 12c must represent the Pourbaix corrosion diagram for gold, with the upper domain representing corrosion as indicated in Figure 48c (if the upper domain represented passivation, the two metals would be of similar stability). Thus, Figure 12a must correspond to the Pourbaix corrosion diagram for rhodium, the regions being as represented in Figure 48a.

Figures 12b and 12d must then be the Pourbaix corrosion diagrams for aluminium and hafnium. As aluminium is used to make cooking utensils for use at pH 7, the region of the Pourbaix corrosion diagram around pH 7 must be immunity or passivation. Hafnium is *less* susceptible to corrosion than aluminium, so the passivation plus immunity regions for hafnium must be greater than those for aluminium. Thus, Figure 12d cannot be the diagram for aluminium (with the region at pH 7 being labelled passive), so Figure 12b must be. This means that Figure 12d is the Pourbaix corrosion diagram for hafnium. The regions must be labelled as shown in Figures 48b and 48d.

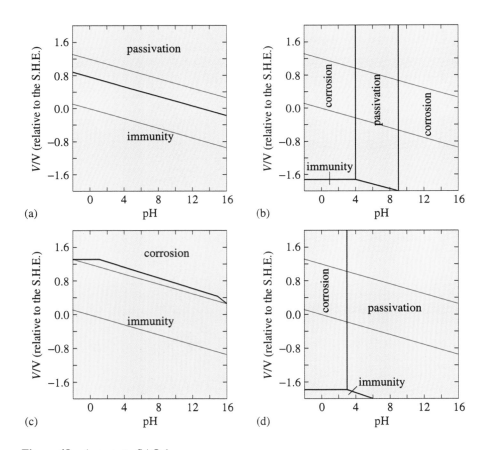

Figure 48 Answer to SAQ 1.

SAQ 2 (Objective 2)

The Evans diagram shown as a plot of V against the logarithm of the current density is shown in Figure 49a.

(i)   If the values for the anodic reaction remain constant, then an increase in the value of E_{ca} to E_{ca}' will produce the effect shown in Figure 49b, increasing the corrosion current density to i_{corr}', and hence increasing the rate of corrosion.

(ii) and (iii)   The effect of increases in the values of $i_{e,\,ca}$ and $\alpha_{ca,\,red}$ are shown in Figures 49c and 49d, respectively. (Remember that the slope $= 2.303RT/\alpha F$, so that increasing α reduces the slope.) Both changes bring about an increase in the value of the corrosion current density, to i_{corr}'' and i_{corr}''', respectively.

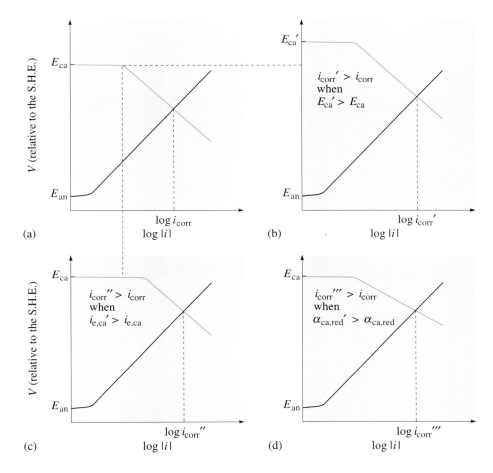

Figure 49   Answer to SAQ 2.

SAQ 3 (Objective 2)

For the zinc reaction

$$Zn^{2+}(aq) + 2e = Zn(s) \qquad\qquad (29)$$

$$E = E^{\ominus} - \left(\frac{RT}{2F}\right)\ln\left\{\frac{a(Zn)}{a(Zn^{2+})}\right\}$$

At 298.15 K, $E^{\ominus} = -0.76$ V; given that $a(Zn) = 1.0$, and neglecting activity coefficients such that:

$a(Zn^{2+}) = c(Zn^{2+})/c^{\ominus}$, with $c(Zn^{2+}) = 10^{-6}$ mol dm^{-3} and $c^{\ominus} = 1$ mol dm^{-3},

then

$$E = -0.76\,\text{V} - \left(\frac{0.059\,2}{2}\,\text{V}\right)\log\left(\frac{1}{10^{-6}}\right)$$

$$= -0.76\,\text{V} - (0.029\,6\,\text{V})\,(6)$$

$$= (-0.76 - 0.178)\,\text{V} = -0.94\,\text{V}$$

Similarly, for the hydrogen ion reduction reaction:

$$E = E^\ominus - (0.059\,2\ \text{V})\,\text{pH} \quad \text{and} \quad E^\ominus = 0.00\ \text{V}$$

So, at pH 2, $E = -0.12$ V.

Thus, the two lines on the Evans diagram will be separated by 0.82 V.

For an α value of 0.5 (hydrogen ion reduction reaction), the slope of the V versus log $|i|$ curve will be -120 mV. With $\alpha = 1.5$ (for the zinc oxidation reaction), the slope will be 40 mV (see Block 8, Table 2). If the areas of anodic and cathodic sites are equal, the horizontal axis of the Evans diagram can be labelled log $|i|$ rather than log $|I|$. If the quoted values of i_e are used, the resulting Evans diagram is as shown in Figure 50. Assumption (b) tells us that $IR_S = 0$, and assumption (c) tells us that the process is not limited by concentration polarization, so the point of intersection of the two lines gives the value of the corrosion potential (relative to the S.H.E.) and the corrosion current density. From the diagram, log $|i_{corr}/\text{A m}^{-2}| \approx 0$, so $i_{corr} = 1$ A m^{-2}.

The assumption likely to be least valid is assumption (f). Under corrosion conditions the concentration of Zn^{2+}(aq) is put at 10^{-6} mol dm^{-3}. It is highly unlikely that the value of the exchange current density (determined at standard concentration conditions, 1.0 mol dm^{-3}) will be unaffected by such a large change in concentration. The assumption will, of course, be somewhat more valid for the hydrogen cathodic reaction at pH 2. (The point is discussed further in Section 3.2.)

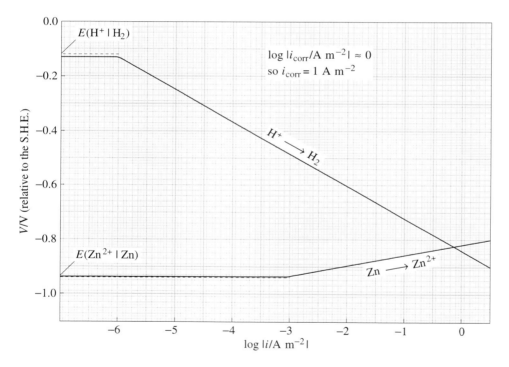

Figure 50   Answer to SAQ 3.

SAQ 4 (Objective 2)

For the nickel reaction

$$Ni^{2+}(aq) + 2e = Ni(s) \qquad (30)$$

with $a(Ni^{2+}) = c(Ni^{2+})/c^\ominus = 10^{-6}$,

$$E(Ni^{2+}|\,Ni) = E^\ominus(Ni^{2+}|\,Ni) + (0.059\,2\ \text{V}/2)\log 10^{-6}$$

$$= (-0.24 - 0.178)\ \text{V} = -0.42\ \text{V}$$

We are told that the value of i_e for nickel formation on nickel under *standard concentration conditions* is 10^{-5} A m^{-2}. If the value is proportional to the concentration of Ni^{2+} (as for iron), then when $c(\text{Ni}^{2+}) = 10^{-6}$ mol dm^{-3}, $i_e = 10^{-5} \times 10^{-6} = 10^{-11}$ A m^{-2}. If the mechanism of nickel oxidation is the same as that for iron, then $\alpha_{\text{an,ox}} = 1.5$, so the Tafel slope is 40 mV.

For the hydrogen cathodic reaction, pH 0 corresponds to standard conditions, so $E(\text{H}^+ \mid \text{H}_2) = 0.0$ V and $i_e = 10^{-1}$ A m^{-2} (from the information given). If $\alpha_{\text{ca,red}} = 0.5$ (as for hydrogen ion reduction on iron), the Tafel slope will be -120 mV.

For the oxygen cathodic reaction, pH 0 again corresponds to standard conditions, so $E = E^{\ominus} = 1.23$ V, and $i_e = 10^{-10}$ A m^{-2} (as for the oxygen cathodic reaction on iron). Again, assuming the same mechanism as on iron, $\alpha_{\text{ca,red}} = 0.5$, so the Tafel slope = -120 mV.

The resulting Evans diagram is shown in Figure 51. This diagram reveals that in unagitated solutions, the rates of the two competing reactions are very similar and both will occur. Thus, $i_{\text{corr}} = i(\text{O}_2) + i(\text{H}_2) \approx 1 \times 10^{-1}$ A m^{-2} + 1×10^{-1} A m^{-2} $\approx 2 \times 10^{-1}$ A m^{-2}. In agitated conditions the oxygen cathodic process will predominate.

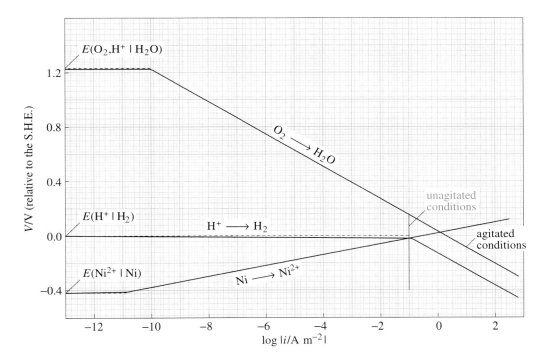

Figure 51 Answer to SAQ 4.

SAQ 5 (Objective 2)

The required Evans diagram for the anodic oxidation of iron (with $E(\text{Fe}^{2+} \mid \text{Fe}) = -0.64$ V, $i_e = 10^{-8}$ A m^{-2} and $\alpha_{\text{an,ox}} = 1.5$) and for the cathodic reduction of hydrogen ions (with $E(\text{H}^+ \mid \text{H}_2) = -0.12$ V, $i_e = 10^{-3}$ A m^{-2} and $\alpha_{\text{ca,red}} = 0.5$) is shown in Figure 52. Assumption (d) tells us that the cathodic reaction is unaffected by hydrogen sulfide ions, so the effect of an increase in the corrosion rate by a factor of 10 can be shown by the double-headed arrow in Figure 52. The Tafel line for the iron anodic reaction can be made to go through this point if either:

(a) i_e is increased to about 10^{-4} A m^{-2} (the value of $\alpha_{\text{an,ox}}$ remaining constant at 1.5 so that the slope is unchanged) — see the full green lines; or

(b) the value of i_e remains constant at 10^{-8} A m^{-2}, the new Tafel line having a slope of about 20 mV (from the Figure — see the broken green line), so, from the relationship $2.303RT/\alpha F = 20 \times 10^{-3}$ V, $\alpha = 3$.

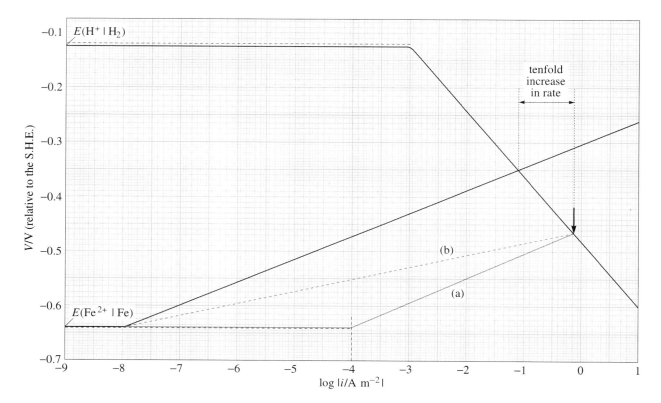

Figure 52   Answer to SAQ 5.

SAQ 6 (Objective 3)

(a) If the drop is saturated with oxygen, then the initial distribution of anodic and cathodic sites will be as before, but this situation will persist for an abnormally long period. However, the final distribution of sites will not be affected.

(b) If the drop is depleted in oxygen, then the initial and final distributions of sites will be the same; that is, reduction of oxygen will occur around the periphery of the drop (where there is greater access to atmospheric oxygen), and oxidation of iron will occur near the centre.

(c) If oxygen is blown into the centre of the drop and nitrogen gas surrounds it, the cathodic reaction will tend to take place at the centre, and the anodic reaction at the periphery; that is, the final distribution of sites is the reverse of the normal situation.

SAQ 7 (Objective 3)

Figure 36   For a partially submerged steel pipe, the area near the surface of the water will be richer in oxygen than the area at greater depths. Thus, the oxygen cathodic reaction will tend to take place near the surface, whereas the anodic reaction, and hence corrosion, will take place lower down.

Figure 37   The sand will be relatively depleted in oxygen compared with the water. Thus, the cathodic reaction will take place on the iron in the water, the anodic reaction being confined to the iron embedded in the sand. Thus, it is the part of the iron embedded in the sand that corrodes preferentially.

Figure 38   The iron in the sand — the soil richer in oxygen — will provide the site of the cathodic reaction, whereas the iron will corrode in the clay, the soil relatively depleted in oxygen.

SAQ 8 (Objective 3)

(a) Slight, copper; (b) slight, solder; (c) severe, cadmium; (d) slight, magnesium; (e) slight, zinc; (f) severe, magnesium (irrespective of whether the stainless steel is active or passive).

SAQ 9 (Objective 3)

For copper connected to cast iron or steel, Table 1 does indeed reveal that corrosion should be moderate, with the cast iron or steel experiencing enhanced corrosion. However, we don't usually circulate seawater, so Table 1 may not be relevant. Also, in central heating systems, the area of the radiator compared with the area of the pipework is usually large, so even though the corrosion current may be substantial, the corrosion current density may be small. In addition, domestic central heating systems tend to be closed systems, so the water in the pipes is not being continuously replaced but is recycled. Thus, when some corrosion has taken place, the oxygen in the water will tend to be used up, so the cathodic reaction will slow down.

ACKNOWLEDGEMENT

Figure 1 from U. R. Evans, *The Rusting of Iron: Causes and Control*, Edward Arnold, 1972 (we have been unable to trace the copyright holder of the photograph, B. J. Stokes).